ORGANOGENESIS OF FLOWERS

ORGANOGENESIS OF FLOWERS

A Photographic Text-Atlas

Rolf Sattler

University of Toronto Press

ISBN 0-8020-1864-5
Microfiche ISBN 0-8020-0193-9
LC 73-185736

MEINEN ELTERN IN

DANKBARKEIT GEWIDMET

CONTENTS

ACKNOWLEDGMENTS

This book could not have been written without the invaluable assistance of students and technicians. I am grateful to all of them. The technicians to whom I am indebted are Mrs Connie Charlton, Mrs Sonia M. Druery, Mrs Ravinder Singh and Mr W. Ziegler. Three former graduate students who have made outstanding contributions are Mrs Vera Block, Mrs Mathilda Cheung and Dr Alastair D. Macdonald. Dr Macdonald and especially Mrs Block carried out their research on floral development in conjunction with this project. They kindly gave me permission to include major portions of their theses. Chapters 1-6 are taken from the PHD thesis by Dr Macdonald entitled 'Floral Development in Some Members of the "Amentiferae." ' Chapters 42-50 are from Mrs Block's MSC thesis 'Floral Organogenesis in Monocotyledons.' These theses were submitted to the Faculty of Graduate Studies and Research of McGill University in 1971 and 1970 respectively. Only very few slight changes and additions were made mainly to adapt the portions of these theses to the format of this book. Dr Macdonald himself kindly wrote the sections on 'Other Authors' specially for this book. Mrs Mathilda Cheung, who completed her MSC thesis on the floral development of *Lythrum salicaria* in 1966, took new photographs of all developmental stages of this species for this book. She also assisted with a few other species. Mr. Usher Posluszny, who is now engaged in a PHD project on floral development in the Potamogetonaceae, made outstanding contributions. He worked on many of the species and succeeded in making many of the most difficult dissections and photographs. The chapter on *Euphorbia splendens* is based on a paper by him which he wrote for a course on 'Special Topics in Floral Development.' Mrs Vera Block worked on several other species besides the ones of Chapters 42-50 with great dedication. A number of former undergraduate students also made fine contributions. Mr Timothy Dickinson did all of the difficult dissections and photography of the chapter on *Pisum sativum*. The following students, most of whom took my course on 'The Flower,' contributed photographs to the chapters whose genus name is indicated in parentheses: R.C. Billick (*Fragaria*), V.E. Cobham (*Chelidonium*), H.C. Gilbey (*Rhus*), F. Gluckman (*Tagetes*), Susan D. Johnston (*Rhus*), Sonia P. Juvik (*Valeriana*), J.C.Y. Law (*Malva*), Jeanette Leung (*Albizia, Pyrola*), D. Liot (*Pelargonium*), Patricia MacNaughton (*Syringa*), Hela Nadler (*Fragaria*), R.A. Proctor (*Hypericum*), C. Tsoukas (*Lantana, Silene*), W.J. Valles (*Tagetes*), B.A. Ziola (*Chelidonium*). Other undergraduate students have contributed indirectly by their valuable work on floral organogenesis. Mr A.F. Muhammad, a graduate student of mine working on vessel structure, took one of the photographs of the chapter on *Scilla*.

I want to thank very much Mrs Eva Krivanek for drawing with great patience and perfection all the floral diagrams, and Miss Stella Sabljak for doing much of the final printing with great care and dedication. To Mrs Evelyn Fung-A-Ling I am grateful for typing the whole manuscript and all of the last minute corrections and additions.

Mrs Judith Farrow read the whole manuscript and suggested improvements in style and content. Dr G.A. Yarranton offered valuable advice on the introduction. Dr A.D. Macdonald and Drs Ronald and

Nancy Dengler also read the introduction and commented on it. Dr V. Singh advised me on the chapters on *Alisma* and *Butomus* and contributed two photographs of *Alisma*. Dr R.I. Greyson and Dr P. Leins commented on the chapters on *Solanum* and *Hypericum* respectively. I want to thank all of them very much. And I also want to thank other colleagues who have helped me in various ways.

I am grateful to the National Research Council of Canada for financial support of the research project through grant no. A2594. The publication of the book was also assisted by a grant from NRC.

In spite of the great support I have received, there have been all sorts of hurdles in the way to the completion of this book. However, eventually all of these hurdles have been passed, I suppose, with the support and inspiration of flower power.

ROLF SATTLER

INTRODUCTION

In 1857, the French botanist Jean-Baptiste Payer published a monumental work on organogenesis of flowers: the famous *Traité d'organogénie comparée de la fleur*. The work consists of two volumes: text and atlas, both of outstanding quality for that time. It is the first and only work of its kind. Since that time the subject has been dealt with in a number of papers and included in several books (e.g., Goebel 1884; Schumann 1890; Church 1908; Jones 1939; Wardlaw 1965), but no other comprehensive work has appeared.

One purpose of this book is to partially fill the void in the literature. It has much in common with Payer's volumes, but it differs in several respects. Whereas Payer documented his observations by drawings, the observations in this work are illustrated by photographs. Thus, this atlas is a photographic atlas. Payer interpreted his observations in terms of a floral theory, but no such attempt is made in this work. With the exception of the floral formulae, this work is meant to be only descriptive. This poses a problem: many of the terms used in floral morphology are interpretive. Examples of such interpretive terms are 'carpel,' 'stamen,' 'ovule,' 'integument,' 'flower.' If such terms are used for the description of floral organogenesis, one describes and interprets at the same time. To give a pure, unbiased description, one needs descriptive terms. Unfortunately, not enough of these are available. Therefore, in this book, some of the interpretive terms are used in a purely descriptive sense (see Glossary of Terms):

for example, 'integument' is used in the sense of 'covering of megasporangium,' not at all as a category fundamentally different from a 'phyllome.' No homologies are implied in descriptive terminology; that is left to the reader. It must be emphasized that even the term 'flower' is used descriptively because a number of flowers included in this book may be interpreted as 'inflorescence,' or 'arillate ovule,' or they may be more homologous to an 'inflorescence' or an 'ovule' than to a 'flower' (Sattler 1966, p. 424; Croizat 1968). 'Carpel primordium' also could be used in a descriptive sense, but, then, in several cases discrepancies would arise between its use in a descriptive and interpretive sense. Therefore, the descriptive term *'gynoecial* primordium' (see Glossary of Terms) is used instead of 'carpel primordium.' And correspondingly in a number of cases the term *'androecial* primordium' has been applied instead of 'stamen primordium.' The concept of 'congenital fusion' was discarded because it does not refer to any fusion that can be observed during ontogeny. Only postgenital fusion can be observed. Therefore, the term 'fusion' is restricted to 'postgenital fusion.'

What is the reason for aiming at a purely descriptive work? As is well known, in comparative plant morphology we find various contradictory conceptual models for the interpretation of the flower (see e.g. Sattler 1965; Meeuse 1966). One such model interprets the flower as a modified monaxial shoot (see e.g. von Bertalanffy 1965; Eames 1961; Cronquist 1968; Takhtajan 1969; Troll 1928, 1957; Zimmermann 1959). According to another model, the basic unit of flowers is the gonophyll (Melville 1962-63). Meeuse (1966) uses the model of an anthocorm and gonoclad for the interpretation of flowers, while Croizat (1960, 1962, 1964) interprets a flower in a less rigid way as an axis with scales some or all of which are sexualized; he distinguishes two types of flowers: the panstrobilar and circumnucellar flowers. Still other conceptual models could be added, each with its own interpretive terminology (see e.g. Meeuse 1966, Chapter 13: 'The outstanding controversy – Floral theories'). In such a situation of conflict, it seems important to focus on what should be the common basis of all of these contradictory conceptions: the description of individual observations, and the description of the entire floral organogenesis which, given the technique employed, must be deduced from the observations of actual developmental stages. It follows that 'pure description' is a relative term. Description of floral organogenesis is not pure in the sense that the actually observed developmental stages have to be combined in the mind of the investigator (Sattler 1962) but it is pure (neutral) in terms of conceptual models of the flower. While a philosophical discussion of 'pure description' is beyond the scope of this book, I want to point out, however, that even 'pure description' of individual observations is a function of the language and concepts used. Kuhn (1962) said: 'As for a pure observation-language, perhaps one will yet be devised ... No current attempt to achieve that end has yet come close to a generally applicable language of pure percepts.' Since very much emphasis is placed on photographs in this book, many, or all, of the difficulties inherent in the description and communication by means of a language are hopefully overcome.

Because of the use of descriptive terminology, the information is more useful to adherents of different conceptual models. It will be also more useful to future botanists who may develop more adequate models of the flower, and to physiologists, geneticists, horticulturists, and other biologists who – for their work – may be interested in the organogenesis of flowers, but not in the interpretation of organogenesis.

At least two kinds of techniques could be used to study floral organogenesis. One is the familiar microtechnique (see e.g.

Johansen 1940; Feder and O'Brien 1968) which involves serial sectioning of floral buds and a subsequent reconstruction of the whole buds from the sections. The other technique is the dissection of floral buds and subsequent photography of the buds (Sattler 1968). Both techniques have advantages and disadvantages. The advantage of sectioning is that it reveals much cytological detail of all tissues. Its disadvantage is the difficulty of reconstructing the three-dimensional picture of the floral buds. The advantage of the dissection technique is that one obtains directly a three-dimensional picture of the floral buds with no reconstruction necessary. However, only the protoderm is visible: inner tissues cannot be studied. It would be ideal to use both techniques in a work like this, but I decided to use only the dissection technique. Since the enormous diversity in floral morphology – which is my main interest – is due to differential growth in developing floral buds, and differential growth occurs in three dimensions, it can therefore be better shown by the dissection technique than by serial sectioning of buds.

It may be objected that the development of internal form – which is omitted here – is more important than the development of external form – which is shown here. Important for what? According to the principle of conservatism of vascular tissue, the vascularization of the flower is more relevant to comparative and phylogenetic interpretations than external form. If this principle is correct, it would be highly desirable to include anatomical data in any comprehensive work. However, it is very questionable whether this principle is correct (see e.g. Hagemann 1963; Esau 1965; Cheung and Sattler 1967; Heel 1969; Carlquist 1969). It seems that in many cases the contradiction of this principle agrees better with developmental observations than the principle itself; at the floral apex growth patterns seem to be correlated with the position of procambial strands; hence a change in growth patterns may change vascularization, and consequently vascularization is not conservative. If this is correct, the study of differential growth, i.e., external form, is important not only for its own sake, but also with regard to anatomy and phylogeny. But regardless of these considerations, the dissection technique demonstrates a key phenomenon which is responsible for diversity in organic form, namely differential growth.

As I pointed out, one purpose of this book is to partially fill a gap of long standing in the botanical literature. Another purpose is to increase our knowledge of floral organogenesis. Since so little is known of floral development this is not too difficult to accomplish. The floral organogenesis of many of the 50 selected species is described here either for the first time or in greater detail than before. The floral organogenesis of other species has been described before by Payer (1857) or by other botanists. But this is the first use of the refined dissection technique (Sattler 1968) which gives a direct picture of the three-dimensional buds, and, at the same time, demonstrates protoderm cells.

Another purpose of this book is to attempt to exert an influence on contemporary botanical thinking, and on the direction in which modern botany and biology are developing. Nowadays in developmental morphology great emphasis is put on causal interrelations. This aspect of enquiry necessarily leads to the study of lower levels of organization: cells, cell organelles, molecules, etc. As a result, cell biology and molecular biology have become very dominant fields of enquiry, whereas comparative studies of organs, whole organisms, and their development have been neglected. This book, with its bibliography, shows how strikingly little is known about the comparative organogenesis of an almost ubiquitous organ-system such as the flow-

er. The floral organogenesis of only 50 species is described here. There are over 200,000 species of flowering plants. Although it may be unnecessary to describe the floral organogenesis of all species of flowering plants, a great many more will have to be studied to get an impression of the diversity in floral organogenesis. Furthermore, it will be important to study the selected 50 species more completely, to investigate the variation within each species more thoroughly and to use additional techniques such as microtechnique, histochemistry, electron microscopy, etc.

The importance of organogenetic and comparative studies is stressed. With regard to the flower, the supposedly typical organogenesis, as mentioned in textbooks, is only one pattern of development among many others which are presented here. Of course, there are many flowers whose development is rather similar to that of a certain type of monaxial shoot, but there are also many other flowers whose development is rather unlike that of the selected shoot. One should not overlook that one of the main characteristics of flowers is diversity. Even if we could gain a full understanding of the pea and sunflower, we would still have to cope with this diversity.

Keeping this in mind, how can the modern morphologist, taxonomist, phylogenist, morphogenesist, and physiologist come to valid general conclusions? On the one hand, by the most detailed study of individual plants or plant species; on the other hand, by a study of as many taxa as possible to grasp the whole range of diversity. As our techniques become more and more refined, there is a danger that we might become preoccupied with the study of a few 'pet plants' and lose sight of the diversity.

Since this book is descriptive, the variation which occurs is not analyzed, and no synthesis of all the data is presented. However, a few concepts may briefly be mentioned which allow us to view the diversity as an expression of a few variables in a common scheme of organization. 'Variation' here is understood in the wide sense, comprising genotypic and phenotypic variation. Some of the most striking kinds of variation, which may be found in the selected 50 species, are the following:

Variation in size and distribution of protoderm cells
In the majority of the photographs the protoderm cells are more or less clearly delimited by their walls. In some of the photographs nuclei and/or nucleoli are visible; but to demonstrate mitotic figures higher magnifications would be necessary. Cell size may vary in different parts of any bud, and in different developmental stages of buds. It also may vary in comparable developmental stages of buds belonging to different taxa. Like cell size, the distribution of protoderm cells is not mentioned in the descriptive part, but the photographs often show it. Since certain correlations in cell arrangement between the cell layers of the apex and primordia can be postulated, some predictions about the arrangement of cells in layers underneath the protoderm may be made.

Variation in the relation between number of cells, size of cells, and size of the floral apex and primordia
Again, these relations are not mentioned in the text, but a glance at the photographs shows that considerable variation occurs in these respects.

Variation in the size of the floral apices, primordia, and whole floral buds
This kind of variation may refer to different developmental stages of buds on one plant (or one plant species), or to comparable developmental stages in different taxa. Variation in growth rates is often conspicuous between primordia of the same bud (e.g. retarded growth of petal primordia),

or primordia in buds of different species.

Variation in the shape of the floral apex and primordia
This kind of variation is related to variation in size. To some extent it may be expressed by ratios of various measurements. But quantification of form is still problematical (see e.g. Schüepp 1965).

Variation in the number of primordia
This kind of variation may refer to the whole flower, or to specific whorls or categories of appendages. Even in species whose numerical plan of the flower is reported as constant, variation was often observed among buds of the same plant or a colony of plants.

Variation in the position of primordia
The position of primordia may be described in terms of phyllotaxis. However, 'phyllotaxis' is a concept which usually refers to the position of primordia on an apex or axis. Since primordia or floral appendages may be formed on primordia of other floral appendages (see e.g. Cheung and Sattler 1967; Sattler 1967), not all observed positions can be characterized in terms of phyllotaxis.

Variation in the sequence of primordial inception
Primordia may be formed in a strictly centripetal sequence, or some of them may be initiated centrifugally (for details see Cheung and Sattler 1967). In borderline cases, which occur rather frequently, it is difficult or impossible to decide whether primordia are initiated simultaneously or successively (see Glossary of Terms under 'simultaneous').

Variation in the plastochron of successive primordia or whorls of primordia
Since 'plastochron' is defined as the time that elapses between the inception of suc-

cessive primordia, it cannot be observed on photographs of developmental stages. But to some extent it may be inferred from differences in the size of successive primordia, unless the growth rates of these primordia differ.

Variation in growth between primordia (interprimordial growth)
If growth occurs between primordia in a radial direction, the result is what is called 'adnation' (a non-developmental term). An example is the 'adnation' of petal and opposite stamen. If growth between primordia occurs in a lateral direction, the result is what is called 'coalescence' (also a non-developmental term). An example is the 'coalescence' of petals. If growth between primordia occurs in both directions, 'perigyny' or 'epigyny' may be the result. Interprimordial growth may be accompanied by and be continuous with growth underneath the primordia, i.e., it may be the interprimordial region of zonal growth (see Cusick 1966). However, growth underneath primordia cannot be demonstrated directly with the technique used here. A girdling primordium (see Glossary of Terms) may be conceived of as the borderline case in which the ratio of primordial and interprimordial growth is one. In all the other cases this ratio is larger than one.

Variation in (postgenital) fusion
Postgenital fusion may occur between almost all kinds of primordia and in varying degrees.

Although the information presented in this book will be most important to the plant morphologist and anatomist, it may be useful and relevant to botanists and biologists working in many other areas of research.

Modern biology, to a great extent, is becoming centered around the study of growth and differentiation. In higher plants, apical meristems and their immediate products are very useful tools for a critical study of these general phenomena. It is in these meristems and their derivatives that cells and groups of cells start to behave differently from each other with respect to growth rates, planes of cell division, internal differentiation, etc. (see e.g. Clowes 1961; Wardlaw 1965; Steward 1968). The photographs in this book do not show internal cellular differentiation, but they do demonstrate a great variety in growth phenomena, including changes in the size and arrangement of protoderm cells. Therefore, they may be important and relevant to the morphogenesist and physiologist who are concerned with the functional interpretation of these events.

Systematists have realized for a long time that a natural system of classification should be based on all available characters (see e.g. Cronquist 1968). With respect to floral characters this implies that characters of all developmental stages of flowers should be taken into consideration. It may be objected that the floral organogenesis of only 50 species is too little to be of much general usefulness. There is no doubt, it would be better to know the organogenesis of 500 species. But even the information on the 50 selected species together with all the other information in the literature should be of some relevance to the systematist. It also may be objected that the selection of the 50 species is not representative. This objection poses a problem: what to one author may be a representative species for a taxon, to another author may not be representative. Now, it should be pointed out that the selection of the 50 species was determined by theoretical and practical considerations. The theoretical consideration of prime importance was to present a diversity of developmental patterns. Therefore, two species may have been chosen from one family or order, if they differed markedly in their organogenesis, whereas several families and orders may have been omitted completely, if they show little difference in their floral organogenesis. The theoretical consideration of secondary importance was to present a great diversity of angiospermous taxa. Thus, the 50 selected species belong to 43 different families and 32 different orders of Angiosperms. Practical considerations were often decisive in this choice. For a study of floral organogenesis much material is needed; all developmental stages must be available; the material should be relatively easy to handle; and so forth. For these reasons some species were included which otherwise may have been substituted by other ones. For example, as Professor Takhtajan pointed out in a personal discussion, it would have been better to include *Myrica cerifera* instead of *Myrica gale*. But material of *Myrica cerifera* was not easily available, whereas material of *Myrica gale* could be collected in large quantities 30 miles away from the McGill campus. Therefore, a description of the floral organogenesis of *Myrica gale* was included, but not that of *Myrica cerifera*. The systematist, who may be critical about such choices, should be reminded perhaps that this book is a beginning, and, above all, a challenge to study in much more detail the variation of floral organogenesis within taxa.

To cope theoretically with the enormous diversity of floral form (and organic form in general) has been a problem of long standing. Recently, numerical taxonomists have made encouraging attempts to create a more objective, repeatable, and quantitative approach to taxonomic problems (Sokal and Sneath 1963). Unfortunately, comparative morphologists and anatomists are still far away from any satisfactory quantification of form. In fact, the majority of comparative morphologists and anatomists are still adhering to qualitative, essentialistic morphological concepts. This is well demonstrated by their definition of 'homology,' one of the most basic concepts in comparative morphology and anatomy.

Even though this book is not a comparative study it may further an awareness that much of the variation in floral development is more or less continuous rather than strictly discontinuous. As a consequence, it may stimulate a search for new semiquantitative or quantitative models of the flower which will be more adequate than the present qualitative or essentialistic concepts (for literature on this topic see Sattler 1966, 1967).

It is impossible to indicate all botanical and biological disciplines to which this book may be relevant, but I may mention that in addition to pure or basic research it may also be relevant to applied research such as agricultural and horticultural research. Many of the species included in this book (e.g. *Pisum sativum, Fuchsia hybrida*) may be of special interest to the agriculturalist or horticulturist.

Finally, this book also may be interesting to the non-scientist: to any person who just likes to look at and enjoy the fascinating forms and shapes of flowers; or to the creative person who tries to transpose this experience into a work of art. As C.P. Snow (1964) pointed out, our society has been suffering for some time from a deep gulf between scientists on one side and literary intellectuals, including artists, on the other side. To bridge this gulf, an attempt must be made at mutual understanding (see, e.g., Ritterbush 1968). The scientist, for his own full personal development as a human being, should take the arts as seriously as science. And the artist might engage himself in the new worlds of science (Huxley 1963). With this in mind, I invite the artist to look at the pictures in this book, to immerse himself in this beautiful world of developing flowers, and to create art out of this new experience. In this way perhaps the book will make some contribution to bridge the frightening gap between our 'two cultures.' And if it does, it will have achieved one of its highest goals.

The floral organogenesis of all of the 50
selected species is described according to
one uniform format: order and family to
which the species belongs (according to
Engler's 12th Syllabus, 1964), Latin and
common name of the species, floral diagram,
floral formula, sequence of primordial
inception, description of floral organogenesis, other authors, bibliography, and photographs, with legends.

Floral diagram
This is 'a cross-section of a (mature) flower
as it would appear if all parts were at the
same level. It might also be thought of as a
sort of aerial view of a flower in diagrammatic form' (Porter 1967). Expressed more
rigorously, the floral diagram is the diagrammatic projection of the flower (parts)
into a plane which is at a right angle to the
long axis of the flower (see also Eichler
1875). Thus, in contrast to the floral formula, the diagram is meant to be empirical,
not theoretical (interpretive).

Symbols used in the floral diagrams

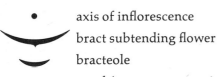

- •   axis of inflorescence
-   bract subtending flower
-   bracteole
-   sepal (or any green perianth member)
-   petal (or any coloured perianth member)
-   stamen with 4 (adaxial) micro-sporangia
-   staminodium
-   pistil without septa, with 1 uni-tegmic ovule (basal placenta-tion) and 2 styles and/or stigmas

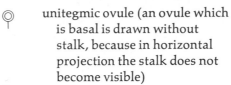

pistil with two septa which are
(postgenitally) fused in the
center; many unitegmic ovules
are inserted on the septa

unitegmic ovule (an ovule which
is basal is drawn without
stalk, because in horizontal
projection the stalk does not
become visible)

bitegmic ovule

median plane

transversal plane

The solid line between the petals indicates
'congenital fusion' which is due to
growth between the primordia.
A dot between two structures indicates
postgenital fusion.
Other symbols used are either self-explanatory or explained in the text.

Floral formula
This is a short symbolic representation of
the mature flower. It describes specifically
the symmetry of the flower, the number
and kind of floral appendages, the number
of whorls, the position of the ovary or ovaries, the number of ovules, and the 'congenital' and postgenital fusion of floral
appendages. The floral formula is interpretive, based on the conception that the flower
is a modified monaxial shoot without axillary buds. Since this idea of the flower is
the predominant one today and usually the
only one dealt with in textbooks, the inclusion of the floral formula may facilitate
understanding of the organogenetic description for those who have received classical botanical training. Those who do not
wish to be biased by the interpretive nature
of the floral formula, and those who are not
familiar with or not interested in the classical interpretation of the flower, should
disregard the floral formula.

Symbols used in the floral formulae

✳ = radial symmetry (disregarding the gynoecium)

·|· = zygomorphic symmetry

+ = bilateral symmetry

P = tepal (perianth member)

K = sepal (calyx member)

C = petal (corolla member)

A = stamen (androecial member)

G = carpel (gynoecial member)

G̲ = superior gynoecium (i.e., flower is hypogynous)

G̅ = inferior gynoecium (i.e., flower is epigynous)

Iv = involucre

D = disc

O = ovule

The number behind a symbol indicates the number of appendages of its kind; e.g., K5 means 5 sepals. ∞ means many (usually more than 12). If the number is put in parentheses this means 'congenital fusion' of the appendages; e.g., K(5) means that the five sepals are congenitally fused at their base. The degree of fusion is not indicated in the formula. Postgenital fusion of appendages is symbolized by parentheses consisting of dotted lines; e.g., K ⦃5⦄ indicates 5 postgenitally fused sepals. Parentheses of solid and dotted lines indicate both 'congenital fusion' and postgenital fusion; e.g., K ⦃(5)⦄ means that the 5 sepals are fused 'congenitally' at their base and postgenitally above. Fusion of appendages of different whorls is indicated by brackets; e.g., [C(5)A5] means that 5 stamens are congenitally fused to a corolla tube; ⸤C(5)A5⸥ would imply postgenital fusion of stamens and corolla tube. If the members of one category are arranged in whorls, the number of each whorl is written separately; e.g., A5+5 refers to an androecium which consists of 2 pentamerous whorls. Thus, the number of whorls (cycles) of the whole flower can be inferred from the floral formula. If within one whorl some members are fused with each other, but not the others,

the fused ones and the free ones are written separately; e.g., A(9)1 means that in one whorl 9 stamens are fused and one is free. Little fusion which is not noticeable in the mature flower is not indicated in the floral formula.

Sequence of primordial inception

In many species primordial inception is centripetal as it is known from shoot apices. But often the sequence of inception of some primordia is reversed. To emphasize the variation in the sequence of inception, it is listed separately in symbolic form. Symbols used for the various kinds of appendages are the same as are used for labelling of the photographs (see below). Symbols for primordia that arise successively are separated by a comma, i.e., all primordia whose symbols are separated by comma(s) appear at the same time. For example, K1,2,3, C1-3 G1-3, O1 would mean that the three sepal primordia (K1,2,3) are initiated in succession, followed by the simultaneous inception of 3 petal (C1-3) and 3 gynoecial (G1–3) primordia, followed by the initiation of one ovule primordium (O1). Not infrequently borderline cases occur for which it is difficult or impossible to decide whether certain primordia are initiated in succession or simultaneously. In such cases, the sequence of primordial inception is accompanied by a note which draws attention to the doubtful portion of the sequence.

Description of floral organogenesis

The description of floral organogenesis starts with the floral apex at a stage of development when it produces the first primordium (or primordia) of the perianth (or androecium or gynoecium in flowers which lack a perianth completely). The inception and early development of the various appendages are described as accurately as possible. Early changes in directions of growth are mentioned, and postgenital fusions are indicated. As the flower reaches maturity, many fine changes in the growth and differentiation of the organs may occur, e.g., petals may become lobed, or indented, or patterns of coloration appear, trichomes and glands differentiate, etc. Such later changes in the development are not described in great detail; in fact, they are often neglected. But in each case the last photograph shows a mature flower. From this photograph and a few others of the later stages of development the reader can get an impression of the slightly neglected advanced stages. Thus, the description of floral organogenesis covers the development of the flower from the floral apex (when it forms the first perianth primordium or primordia) to the mature flower. However, emphasis is put on the early stages of organogenesis when the general pattern of floral construction is determined.

The development of the inflorescence is not included. However, in some cases the classical morphologist might debate whether a certain structure represents a 'flower' or an 'inflorescence.' Such structures which cannot be classified because they show characters of both a 'flower' and an 'inflorescence' are considered as flowers in this book. This does not mean that they are homologized with flowers; they are included simply for practical reasons. As there are intermediates between an 'ovule' and a 'pistil,' intermediates occur between a 'flower' and an 'inflorescence' (see, e.g., Croizat 1968).

The entire description of floral organogenesis is very brief. Its purpose is to orient the reader and give him a basis for a better understanding of the photographs. There is much information in the photographs which is not mentioned in the text. Therefore, the photographs should be considered as the most important part of this book.

Other authors
In this section observations (not interpre-

tations in terms of floral models) of other authors are mentioned, especially if their observations contradict the ones reported here. The discussion of the observations is not intended to be complete. Those who are interested in the detailed organogenetic description given by other authors are referred to the bibliography.

Bibliography
The complete recent literature which deals with the floral organogenesis of the species under consideration is included. The reference for older works is usually given only if it is not quoted in the recent ones. Thus, by means of the given literature and its references the reader should get the complete bibliography on the subject. Often recent references on the floral organogenesis of related species or even related genera are included. In spite of this frequent inclusion of related taxa, it is surprising to see how little work has been done on the majority of the 50 selected species and genera.

Photographs
For each species the photographs of the various developmental stages are numbered, starting with the youngest stage and ending with the oldest stage, the mature flower. Usually a photograph with a lower number is a younger stage than another photograph of the same species with a higher number. For example, figure 6 would be a younger stage than figure 7 (or the same stage as figure 7, if more than one photograph is reproduced of the same developmental stage). But sometimes the development of one kind of organ is shown after that of another kind; for example, first the complete development of the androecium is shown, and then that of the gynoecium. In these cases, an older stage or stages (e.g., of the androecium) may precede a younger stage or stages (e.g., of the gynoecium). Usually, the top views of

floral buds are oriented so that they correspond in position to the floral diagram. Most of the photographs are labelled, using the abbreviations given below. Labels were omitted whenever the photographs seemed to be self-explanatory so that no detail of the photograph would be obscured. In the legend each photograph is described very briefly, and the magnification is indicated. The great majority of photographs are magnified x 146. This facilitates comparison between different species considerably. Other photographs have a magnification of x 85 or rarely of x 246; and the photographs of the oldest stages and mature flowers have a variety of magnifications depending on their size.

Abbreviations for floral structures
A $=$ androecial member or its primordium
A_c $=$ antepetalous androecial member or its primordium
A_k $=$ antesepalous androecial member or its primordium
A_i $=$ androecial member or its primordium of the inner whorl
A_o $=$ androecial member or its primordium of the outer whorl
B $=$ bract or bract primordium
b $=$ bracteole or bracteole primordium
C $=$ petal or petal primordium
D $=$ disc
F $=$ floral apex
Fn $=$ funiculus
G $=$ gynoecial member or its primordium
I $=$ integument or its primordium
I_i $=$ inner integument or its primordium
I_o $=$ outer integument or its primordium
Lo $=$ locule
K $=$ sepal or sepal primordium
N $=$ nucellus (or its primordium) (megasporangium)
O $=$ ovule or ovule primordium
Oa $=$ ovary or developing ovary
P $=$ perianth member or its primordium
Pl $=$ placenta or placental primordium

R $=$ inflorescent apex (reproductive apex)
r $=$ removed, e.g., rA $=$ stamen removed
Se $=$ septum or developing septum
Si $=$ stigma or developing stigma
Sp $=$ spur or developing spur
Sy $=$ style or developing style
T $=$ theca or its primordium
Abbreviations which are used infrequently are explained in the text.

GLOSSARY OF TERMS

Abaxial A position on the median plane (of a flower or floral bud) between the transversal plane and the subtending bract of the flower or floral bud. A structure is said to be abaxial when its median plane coincides with the median plane of the flower under the above condition. 'Towards the abaxial side (of a flower or floral bud)' refers to a position near the abaxial position somewhere between an abaxial and transversal position. For a diagrammatic explanation of 'abaxial' see under 'Symbols used in the floral diagrams.'

Adaxial A position on the median plane (of a flower or floral bud) between the transversal plane and the inflorescence axis. A structure is said to be adaxial when its median plane coincides with the median plane of the flower under the above condition. 'Towards the adaxial side (of a flower or floral bud)' refers to a position near the adaxial position somewhere between an adaxial and transversal position. For a diagrammatic explanation of 'adaxial' see under 'Symbols used in the floral diagrams.'

Androecial Primordium A primordium which develops into an androecial member, i.e., a stamen, part of a stamen, or a group of stamens (or a branched stamen).

Antepetalous A position opposite (superposed to) the petal or petal primordium (e.g., the stamens in *Lysimachia* have an antepetalous position).

Antesepalous Opposite sepal.

Antetepalous Opposite tepal (e.g., the stamens in *Comandra* are antetepalous).

Floral Apex That portion of a floral bud inside the innermost primordium or primordia. It is usually impossible to delimit the floral apex sharply from the surrounding primordium or primordia.

Floral Bud A young receptacle with a primordium or primordia of floral appendages. Even very young stages in which the primordia have not yet overarched the floral apex are called floral buds.

Floral Organogenesis Inception and development of the floral organs. In this book 'floral organogenesis' refers to the development of the external form of the flower.

Flower A receptacle with fertile (and sterile) appendages. In developmental terms: the structure that develops from a floral apex and the primordia which it produced. Note: since a primordium of an appendage is not necessarily fundamentally different from a floral apex, intermediates between a primordium and a floral apex occur; consequently, intermediates between a flower and an inflorescence occur. The definition of 'flower' given here does not take into consideration such intermediates. Therefore it is of restricted usefulness, and often arbitrary decisions are necessary as to what constitutes a flower. Some flowers may lack functional fertile appendages.

Fusion Unless specified otherwise, 'fusion' is used in the sense of postgenital fusion, i.e., fusion observable during ontogeny.

Girdling Primordium A primordium that encircles completely the floral apex (or any other portion of the developing flower). Its outline may be circular (circular primordium), or elliptical (elliptical primordium), or with three, four, five, etc. corners (i.e., trigonal, tetragonal, pentagonal, etc. primordium). Since a girdling primordium surrounds completely the floral apex, it may be called a ridge or rim around the floral apex. This rim or ridge is of equal height throughout its circumference. Often primordia arise on such a girdling primordium. Borderline cases occur rather frequently between a girdling primordium and a whorl of primordia 'connected' by interprimordial growth.

Gynoecial Primordium A normal or girdling primordium which forms the whole gynoecium or part of it, excluding the placenta(e) and/or ovule(s) which develop from placenta and ovule primordia. The primordia of ovules (and placentae) are formed on the gynoecial primordia or other floral tissue. **Generally**, a gynoecial primordium develops into a gynoecial member which may be a whole pistil or part of a pistil. Note: in some cases a pistil develops from one girdling primordium and normal primordia which are formed on the girdling primordium. To say that 'two gynoecial primordia are formed on the girdling gynoecial primordium' is the same as saying that 'the girdling gynoecial primordium becomes two-lobed.'

Interprimordial Growth Growth between primordia. Interprimordial growth may occur, for example, between the primordia of one whorl or between superposed primordia of different whorls. Interprimordial growth may be combined and continuous with growth underneath the primordia, i.e., it may be part of zonal growth (see Cusick 1966). With the technique employed in this book, growth underneath primordia cannot be detected; therefore the phrases 'growth between primordia' and 'interprimordial growth' do not exclude 'zonal growth': they refer to growth between primordia and to zonal growth which includes regions between primordia. (See also p. xv.)

Median Plane A longitudinal plane going through the center of the floral apex and the subtending bract of the flower or floral bud. For a diagrammatic explanation see under 'Symbols used in the floral diagrams.'

Opposite On the same radius and on the same side of the floral bud or flower. (Superposed is used as a synonym.)

Primordium A growth center formed on the periphery of the (floral) apex or another primordium. Usually a primordium cannot be sharply delimited from the floral apex and the developing receptacle (see also under 'Floral Apex,' and Sattler 1967). The floral apex itself could be called a primordium but in order to avoid confusion 'primordium' is not used for the floral apex.

Postgenital Fusion See 'Fusion.'

Receptacle The axis of a flower which cannot be sharply delimited from the floral appendages.

Simultaneous Inception of Primordia The primordia become visible at the same time. 'At the same time' is a relative term dependent on the degree of accuracy; for example, one observer may say that five primordia arose simultaneously during the last hour. Another observer may claim that these five primordia were initiated successively in 10-minute intervals. Hence, whatever appears simultaneous may turn out to be a succession if more accurate time measurements are taken. Besides that, the inception of primordia – with the technique used in this work – was not observed in time, but deduced from a series of successive developmental stages. Therefore, the statement 'simultaneous inception of primordia' has to be understood with qualifications. (See also 'Successive.')

Spiral Sequence In many cases primordia (e.g., sepal primordia) are said to arise 'in a spiral sequence.' This statement refers to spiral phyllotaxis (e.g., 2/5 phyllotaxis)

implying successive inception of the primordia. If the successive primordia are inserted at slightly different levels, the connection of the primordia by an ideal line yields a spiral. Whether the primordia are inserted at different levels is often difficult or impossible to decide. Therefore, 'spiral sequence' is only defined by the sequence of inception, not by differences in the level of inception, although both properties are usually correlated.

Successive Inception of Primordia Unless all successive developmental stages were found, succession was deduced from differences in the size of primordia. This deduction is permissible if the primordia grow at the same rate; otherwise it may lead to false conclusions. For example, the petal primordia often grow more slowly than the stamen primordia; consequently they may be smaller than the stamen primordia, although they were initiated before. In this case, a deduction of the sequence of primordial inception from size differences would lead to a false conclusion.

Transversal Plane A plane at right angles to the median plane going through the center of the floral apex or flower.

Tube A cylindrical portion. Examples of tubes are *calyx tube, corolla tube, perianth tube, androecial tube*. The gynoecial tube is called the ovary wall or pistil wall. A tube is named according to the category of primordia at whose base it is formed. If other categories of primordia are formed later on a tube, this does not change the name of the tube; e.g., in *Lythrum* after the inception of a calyx tube, epicalyx and corolla primordia are formed on the calyx tube; in spite of this, the tube in *Lythrum* is called a calyx tube. Whether a tube is formed by the 'congenital fusion' of appendages among themselves or with the receptacle is not necessarily a meaningful alternative. In contrast, a tube may be formed by postgenital fusion of appendages: in that case it is clearly appendicular.

Other terms which are used in the text are either self-explanatory, or they are defined when they are used, or they are used in the same sense as defined by Esau (1961), Porter (1967), or Usher (1966).

MATERIALS AND METHODS

The plant material was collected either at the natural habitats of the species or in the greenhouses (and growth chambers) of the Department of Botany at McGill University. In the case of the naturally occurring species, floral buds or young inflorescences were collected in intervals from the same colony or population of plants. Usually all the buds were taken from a small number of plants which grew within a few square yards. The buds of arboreal species often were collected on one tree only. This sampling technique should guarantee that the buds were genetically homogeneous or even identical. The localities, at which the material was collected, are within a radius of 30 miles from downtown Montreal. The material of *Hibbertia scandens* was collected on the campus of the University of California at Santa Barbara, some of it by Dr Katherine Esau. I want to express my gratitude to Dr Esau for this collection.

The plants utilized from the greenhouses of McGill University were grown under average greenhouse conditions. Only *Hordeum vulgare* was grown under special conditions in growth chambers through the courtesy of Dr W.G. Boll. Illumination was continuous by incandescent lights (40 watt bulbs) of about 5.4×10^5 ergs/cm²/sec, and fluorescent lights (GE cool white) of about 3.6×10^5 ergs/cm²/sec. The temperature cycle was 18 hours at 21°c and 6 hours at 16°c. The relative humidity was about 75 per cent.

Immediately after the buds had been re-moved from the plants, they were fixed and preserved in FAA (formalin-acetic acid-alcohol). They were then stained in alcoholic acid fuchsin, dissected, if necessary, and photographed completely immersed in 100 per cent ethyl alcohol (Sattler 1968). Most of the buds which are reproduced are preserved in 100 per cent ethyl alcohol, and they and others have been retained as voucher specimens.

Herbarium vouchers were also prepared from the material studied.

SYSTEMATIC POSITION OF THE 50 SPECIES

The system of classification of Engler's 12th Syllabus (1964) has been used to indicate the systematic position of the 50 selected species. In this Syllabus the Dicotyledons are divided into 37 plus 11 orders (Reihen), and the Monocotyledons contain 14 orders. Most of the orders comprise suborders, families, tribes, etc. In the following list, all orders of the Syllabus are mentioned. The number of suborders and families in each order is listed. Only those families are listed to which the 50 selected species belong. This should give the reader some indication of how the selected species and families are distributed among the families, suborders and orders of Engler's 12th Syllabus.

Dicotyledoneae

Casuarinales / 1 family
Juglandales / 2 families
   Myricaceae: *Myrica gale*
   Juglandaceae: *Juglans cinerea*
Balanopales / 1 family
Leitneriales / 2 families
Salicales / 1 family
   Salicaceae: *Populus tremuloides*
Fagales / 2 families
   Betulaceae: *Ostrya virginiana*
   Fagaceae: *Quercus rubra*

Urticales / 5 families
   Urticaceae: *Laportea canadensis*
Proteales / 1 family
Santalales / 2 suborders and 7 families
   Santalaceae: *Comandra umbellata*
Balanophorales / 1 family
Medusandrales / 1 family
Polygonales / 1 family
   Polygonaceae: *Fagopyrum sagittatum*
Centrospermae / 4 suborders and 13
    families
   Caryophyllaceae: *Silene cucubalus*
   Chenopodiaceae: *Chenopodium album*
Cactales / 1 family
Magnoliales / 22 families
Ranunculales / 2 suborders and 7 families
   Ranunculaceae: *Ranunculus acris*
Piperales / 4 families
   Piperaceae: *Peperomia caperata*
Aristolochiales / 3 families
Guttiferales / 4 suborders and 16 families
   Dilleniaceae: *Hibbertia scandens*
   Guttiferae: *Hypericum perforatum*
Sarraceniales / 3 families
Papaverales / 4 suborders and 6 families
   Papaveraceae: *Chelidonium majus*
   Cruciferae: *Cheiranthus cheiri*
Batales / 1 family
Rosales / 4 suborders and 19 families
   Rosaceae: *Fragaria vesca*
   Leguminosae: *Albizia lophanta, Pisum*
     sativum
Hydrostachyales / 1 family
Podostemales / 1 family
Geraniales / 3 suborders and 9 families
   Geraniaceae: *Pelargonium zonale*
   Euphorbiaceae: *Euphorbia splendens*
Rutales / 3 suborders and 12 families
Sapindales / 4 suborders and 10 families
   Anacardiaceae: *Rhus typhina*
Julianales / 1 family
Celastrales / 3 suborders and 13 families
Rhamnales / 3 families
   Rhamnaceae: *Rhamnus cathartica*
Malvales / 4 suborders and 7 families
   Malvaceae: *Malva neglecta, Alcaea rosea*
Thymelaeales / 5 families

Violales / 5 suborders and 20 families
Cucurbitales / 1 family
Myrtiflorae / 3 suborders and 17 families
   Lythraceae: *Lythrum salicaria*
   Onagraceae: *Fuchsia hybrida*
Umbelliflorae / 7 families
   Umbelliferae: *Anthriscus sylvestris*
Diapensiales / 1 family
Ericales / 5 families
   Pyrolaceae: *Pyrola elliptica*
Primulales / 3 families
   Primulaceae: *Lysimachia nummularia*
Plumbaginales : 1 family
Ebenales / 2 suborders and 7 families
Oleales / 1 family
   Oleaceae: *Syringa vulgaris*
Gentianales / 7 families
   Asclepiaceae: *Asclepias syriaca*
Tubiflorae / 6 suborders and 26 families
   Convolvulaceae: *Calystegia sepium*
   Verbenaceae: *Lantana camara*
   Solanaceae: *Solanum dulcamara*
Plantaginales / 1 family
Dipsacales / 4 families
   Valerianaceae: *Valeriana officinalis*
Campanulales / 8 families
   Stylidiaceae: *Stylidium adnatum*
   Compositae: *Tragopogon pratensis,*
     Tagetes patula

Monocotyledoneae

Helobiae / 4 suborders and 9 families
   Alistmataceae: *Alisma triviale*
   Butomaceae: *Butomus umbellatus*
Triuridales / 1 family
Liliflorae / 5 suborders and 17 families
   Liliaceae: *Allium neapolitanum, Ruscus*
     hypoglossum, Scilla violacea
Juncales / 2 families
Bromeliales / 1 family
Commelinales / 4 suborders and 8 families
Graminales / 1 family
   Gramineae: *Hordeum vulgare*
Principes / 1 family
Synanthae / 1 family
Spathiflorae / 2 families

Araceae: *Acorus calamus*
Pandanales / 3 families
   Sparganiaceae: *Sparganium eurycarpum*
Cyperales / 1 family
   Cyperaceae: *Scirpus validus, Cyperus esculentus*
Scitamineae / 5 families
Microspermae / 1 family
   Orchidaceae: *Habenaria clavellata*

LITERATURE CITED

Bertalanffy, L. von. 1965. Zur Geschichte theoretischer Modelle in der Biologie. Studium Geneale, *18*: 290-298.

Carlquist, S. 1969. Toward acceptable evolutionary interpretations of floral anatomy. Phytomorphology, *19*: 332-362.

Cheung, M. and Sattler, R. 1967. Early floral development of *Lythrum salicaria*. Can. J. Bot., *45*: 1609-1618.

Church, A. 1908. *Types of floral mechanism.* Oxford: Clarendon Press.

Clowes, F.A.L. 1961. *Apical meristems.* Botanical Monographs Vol. 2. Oxford: Blackwell Scientific Publications.

Croizat, L. 1960. *Principia Botanica.* 2 vols. Codicate.

– 1962. *Space, time, form: the biological synthesis.* Caracas.

– 1964. Thoughts on high systematics, phylogeny and floral morphogeny, with a note on the origin of the Angiosperms. Candollea, *19*: 17-96.

– 1968. The biogeography of the tropical lands and islands east of Suez-Madagascar: with particular reference to the dispersal and form-making of *Ficus* L., and different other vegetal and animal groups. Atti Dell'Istituto Botan. Labor. Crittogam. Dell'Univ. Pavia, Serie 6, vol. *4*: 1-400.

Cronquist, A. 1968. *The evolution and classification of flowering plants.* Boston: Houghton Mifflin Co.

Cusick, F. 1966. On phylogenetic and ontogenetic fusions. In: Cutter, E. (ed.). *Trends in plant morphogenesis.* London: Longmans.

Eames, A.J. 1961. *Morphology of the Angiosperms.* New York: McGraw-Hill.

Eichler, A. 1875. *Blüthendiagramme.* Vol. 1. Leipzig.

Engler's Syllabus der Pflanzenfamilien. 1964. 12th ed. H. Melchior. Vol. 2. Berlin-Nikolasee: Borntraeger.

Esau, K. 1961. *Anatomy of seed plants.* New York: J. Wiley.

– 1965. *Vascular differentiation in plants.* Biology Studies. New York: Holt, Rinehart and Winston.

Feder, N. and O'Brien, T.P. 1968. Plant microtechnique: Some principles and new methods. Amer. J. Bot. *55*: 123-142.

Goebel, K. 1884. *Vergleichende Entwicklungsgeschichte der Pflanzenorgane.* Berlin: R. Friedländer & Sohn.

Hagemann, W. 1963. Die morphologische Sproß-Differenzierung und die Anordnung des Leitgewebes. Ber. Deut. Botan. Ges., *76*: 113-120.

Heel, W.A. van. 1969. The synangial nature of pollen sacs on the strength of 'congenital fusion' and 'conservatism of the vascular bundle system' w. sp. ref. to some Malvales, I. Proc. Koninkl. Nederl. Akad. Wetensch. (Amsterdam), Ser. C, *72*: 172-206.

Huxley, A. 1963. *Literature and science.* London: Chatto and Windus.

Johansen, D.A. 1940. *Plant microtechnique.* New York: McGraw-Hill.

Jones, S.G. 1939. *Introduction to floral mechanism.* London: Blackie & Son Ltd.

Kuhn, T.S. 1962-70. *The structure of scientific revolutions.* Chicago: Chicago Univ. Press.

Meeuse, A.D.J. 1966. *Fundamentals of phytomorphology.* New York: Ronald Press.

Melville, R. 1962-63. A new theory of the angiosperm flower. I. The gynoecium. II. The androecium. Kew Bull.

Roy. Botan. Gardens, *16*: 1-50, and *17*: 1-63.

Payer, J.B. 1857. *Traité d'organogénie comparée de la fleur. Texte et Atlas.* Paris: Librairie de Victor Masson.

Porter, C.L. 1967. *Taxonomy of flowering plants.* 2nd ed. San Francisco: Freeman and Co.

Ritterbush, P.C. 1968. *The art of organic forms.* Washington: Smithsonian Institute Press.

Sattler, R. 1962. Zur frühen Infloreszenz- und Blütenentwicklung der Primulales ... Botan. Jahrb., *81*: 358-396.

– 1965. 'Flower.' In: *McGraw-Hill Yearbook Sci. and Technol.* New York: McGraw-Hill Book Co.

– 1966. Towards a more adequate approach to comparative morphology. Phytomorphology, *16*: 417-429.

– 1967. Petal inception and the problem of pattern detection. J. Theoret. Biol., *17*: 31-39.

– 1968. A technique for the study of floral development. Can. J. Bot., *46*: 720-722.

Schüepp, O. 1965. *Meristeme.* Basel: Birkhäuser Verlag.

Schumann, K. 1890. *Neue Untersuchungen zum Blüthenanschluß.* Leipzig.

Snow, C.P. 1964. *The two cultures and a second look.* Toronto: Mentor Books. Reprint of a hardcover ed. by Cambridge Univ. Press, 1959-63.

Steward, F.C. 1968. *Growth and organization in plants.* Reading: Addison-Wesley Publ. Co.

Takhtajan, A. 1969. *Flowering plants: origin and dispersal.* Edinburgh: Oliver and Boyd. (Translated from the Russian by C. Jeffrey.)

Troll, W. 1928. *Organisation und Gestalt im Bereich der Blüte.* Berlin.

– 1957. Praktische Einführung in die Pflanzenmorphologie. 2. Teil. Jena: VEB G. Fischer Verlag.

Usher, G. 1966. *A dictionary of botany.* London: Constable.

Wardlaw, C.W. 1965. *Organization and evolution in plants.* London: Longmans.

– 1965. The organization of the shoot apex. In: Encyclopedia of Plant Physiol., *15*(1): 966-1076.

Zimmerman, W. 1959. *Die Phylogenie der Pflanzen.* 2nd ed. Stuttgart: Fischer Verlag.

DESCRIPTION OF THE FLORAL ORGANOGENESIS OF 50 SPECIES

DICOTYLEDONS / 3

MONOCOTYLEDONS / 168

1 / JUGLANDALES

MYRICACEAE

Myrica gale L. (bog myrtle)

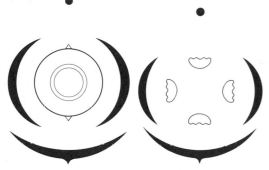

Female flower

Floral formula:
$+ P_2 \bar{G}(2) O_1$
Note: P2 refers to the
two bracteoles.

Male flower
Note: The two
bracteoles (perianth
members) are
inconspicuous in the
mature flower.

Floral formula: $+ P_2 A_4$

Sequence of primordial inception
Female flower: P_{1-2}, $G_{girdling}$, $G_{1-2(3)}$, O_1
Male flower: P_{1-2}, A_{1-4}

DESCRIPTION OF FLORAL ORGANOGENESIS

Female flower
The floral apex arises as an ovoid structure
in the axil of a bract (1). Two bracteole
primordia are formed at the transversal
poles. These become more prominent as
growth occurs evenly at the periphery of
the floral apex causing the apex to flatten
(1). This peripheral growth continues with
the formation of a gynoecial primordium
that girdles the floral apex (2). One of two

Female flower
1 Side view of the distal region of the inflorescence
with bracts removed. The bracteoles are positioned
on either side of the floral apex in the transversal
plane. x 146
2 Top view of a number of floral buds showing the
girdling gynoecial primordium. x 146

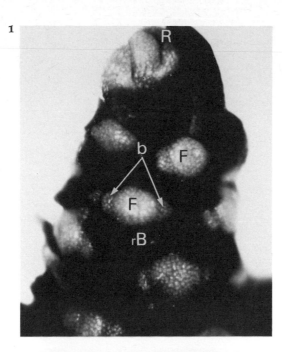

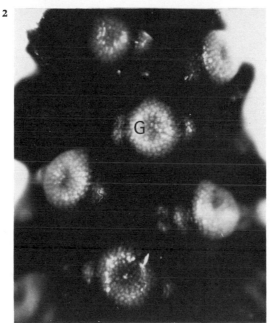

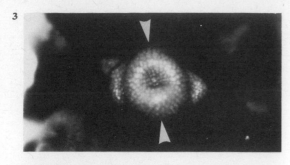

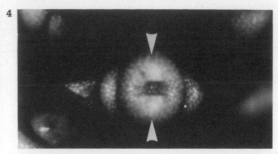

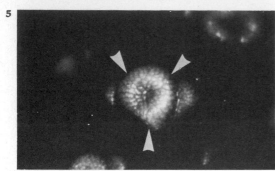

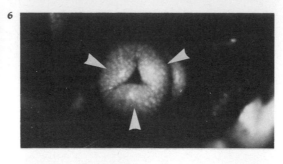

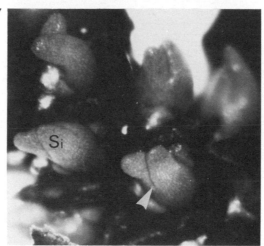

types of growth patterns may now occur on the distal surface of the gynoecial primordium. Two humps may arise simultaneously in the median plane alternate to the bracteoles (3) or less commonly, three humps may form at positions midway between the transversal and median planes (5). These humps, distal to the girdling gynoecial primordium, form the two (4, 7) or the three stigmas (6, 8).

Growth occurs in a zone within the gynoecial primordium so that the ovary wall extends upward as a cylindrical structure carrying with it the bracteoles (7, 8). This cylindrical extensive growth is more active in the region above the bracteoles (8). The ovarian cavity is closed distally by the fusion of the margins of the ovary wall as the stigmas become appressed (7, 8).

At the base of the locule, in the center, the floral apex is transformed into a columnar ovular primordium which is slightly wider in the median plane than it is in the transversal plane (9). The single integument arises as a girdling primordium about the ovular apex (10, 11). Viewed from the transversal plane the integument is slightly two-lobed (10). The integument overgrows the nucellus (12, 13). The distal margin of the integument is slightly convoluted and the base of the integument, where it joins the funiculus, is corrugated (13).

Male flower
The floral apex arises in the axil of a bract. Two bracteole primordia are formed at the transversal poles of the floral apex (14). The bracteoles terminate growth at an early stage (19, 21). The floral apex flattens (15) and peripheral growth activity then occurs at four positions on the flanks of the apex thereby initiating the four stamen primordia (16). Two stamens are formed in the median plane and two are formed in the transversal plane opposite the bracteoles. The floral apex and the four primordial stamens continue to increase in size at an

equal rate (19, 18). The stamens grow slightly outward and dichotomize at their upper distal surface (19) forming two thecae (20, 22). Each theca forms two pollen sacs (20). Growth at the floral apex resumes briefly forming a small hump (19, 21) which does not develop further.

OTHER AUTHORS

Baillon's (1876-79) brief description of floral organogenesis is confirmed by my observations with the exception of gynoecial inception for which he does not mention a girdling primordium but two primordia that become interconnected later on. Chevalier's (1901) study is unfortunately deficient in diagrams. He does not mention the occurrence of bracteoles in the male flower but more significantly he states that the stamen primordia are initiated in a sequence suggesting a spiral relationship; this was not observed in my material. His observations on the development of the female flower agree with mine as does the study by Vikhireva (1957). The latter author, however, does not examine the early stages of floral development nor does he examine

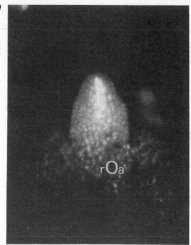

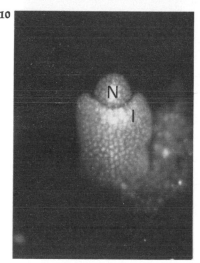

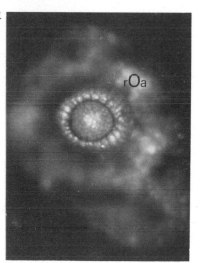

-4 Top view of two floral buds at different stages of growth. The arrowheads point to the two regions on the girdling gynoecial primordium that form the stigmas. x 146

-6 Top views of two floral buds at different stages of growth. The arrowheads point to the three regions that will form stigmas. x 146

7 A number of flowers showing the appression of the young stigmas. x 85

8 Female flowers with two and three stigmas at the time of pollination. x 85

9 Side view of the primordial ovule prior to integument inception. Notice its asymmetrical appearance. The young ovary wall was removed (rOa). x 146

o Side view of the ovule showing the bilobed integument. x 146

1 Top view of the young ovule just after integument inception. x 146

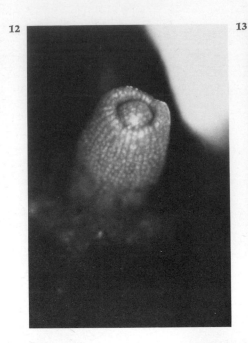

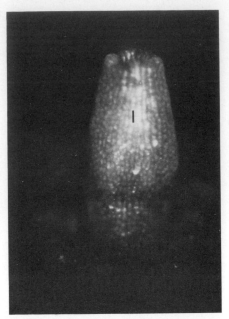

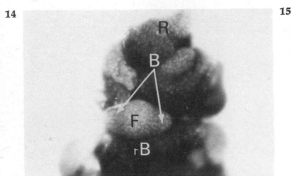

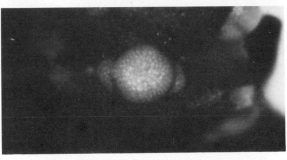

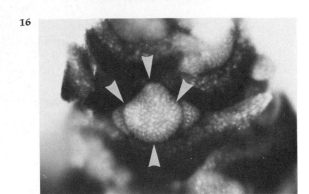

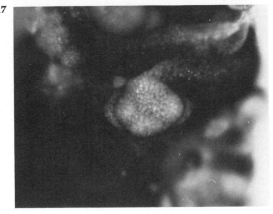

floral development in the male flower. Yen (1950) examines in detail the development of the female flower (gynoecium plus succulent papillae) of *Myrica rubra*. Organogenesis in the female flower of *Myrica gale* and *Myrica rubra* could be interpreted as being similar. Hagerup (1934) records that the integument of *M. rubra* is initiated first on one side of the primordial ovular stalk and then develops around the stalk to fuse postgenitally. This was not observed in my material.

Most studies have simply described the mature flower or fruit. None of the studies of the male flower mention the presence of bracteoles (Eichler 1878; Engler 1894; Hjelmquist 1948; Youngken 1920). With the addition of Kershaw (1909) the descriptions of the female flower by the above authors agree with mine.

BIBLIOGRAPHY

Baillon, H. 1876-79. Traité du développement de la fleur et du fruit. Adansonia 12: 1-19.

Benson, Margaret, and Welsford, E.J. 1909. The morphology of the ovule and female flower of *Juglans regia* and of a few allied genera. Ann. Bot. 23: 625-633.

Chevalier, A. 1901-02. Monographie des Myricacées, anatomie et histologie, classification et description des espèces, distribution géographique. Me. Soc. Sci. Nat. Cherbourg, 32: 85-340.

Eichler, A.W. 1878. *Blüthendiagramme*. Leipzig.

Engler, A. and Prantl, K. 1894. *Die natürlichen Pflanzenfamilien*. Leipzig.

Hagerup, O. 1934. Zur Abstammung einiger Angiospermen durch Gnetales und Coniferae. Biol. Medd. udg. af Danske Vidensk. Selsk. 11: 4, Köbenhavn.

Hjelmquist, H. 1948. Studies on the floral morphology and phylogeny of the Amentiferae. Bot. Notiser, Suppl. 2: 1-171.

Kershaw, E.M. 1909. The structure and development of the ovule of *Myrica gale*. Ann. Bot. *23*: 353.

Vikhireva, V.V. 1957. Anatomical structure and development of the female flower of the common bog myrtle – *Myrica gale* L. USSR Akad. Sci. Series 7 No. 4 270 (In Russian).

Yen, T. 1950. Structure and development of the flower and the fruit of *Myrica rubra*. Peking Nat. Hist. Bull. *19* (1): 2-20.

Youngken, H.W. 1920. The comparative morphology, taxonomy and distribution of the Myricaceae of the Eastern United States. Contrib. Bot. Lab. of the U. of Penn. *4* (2): 339-400.

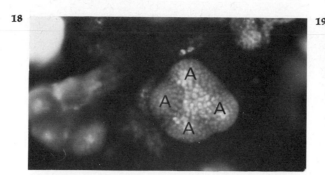

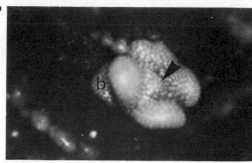

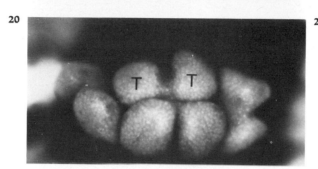

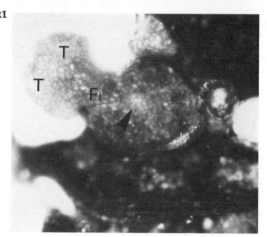

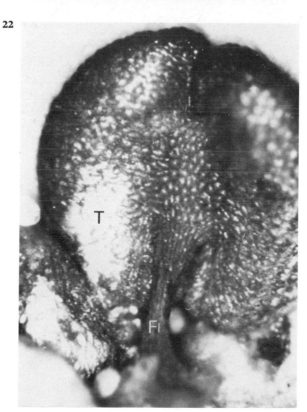

13 Ovules showing different stages of integument development.

Male flower

14 Side view of the distal region of a young inflorescence with bracts removed to show the floral apex and two bracteoles. x 146

18 Top views of floral buds. Successive stages in the inception of four stamen primordia. x 146

19 Top view of floral bud showing the dichotomy of the stamen primordia and the floral apical hump in the center (marked by black arrowhead). x 146

20 Top view of young stamens. The two pollen sacs per theca are evident. x 146

21 Three of the four stamens have been dissected away showing the floral apical hump. x 146

22 Abaxial view of a stamen at the time of anthesis showing the connection of the filament and the anther. x 85

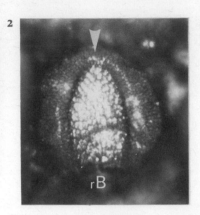

B
F
R

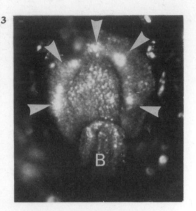

rB

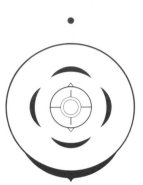

B

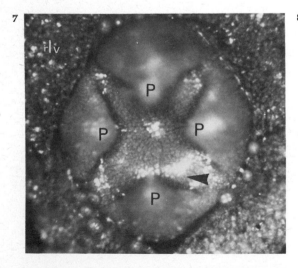

Iv

rB

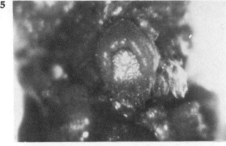

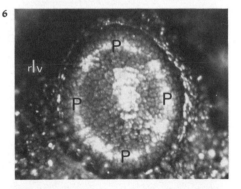

rIv

P
P
P
P

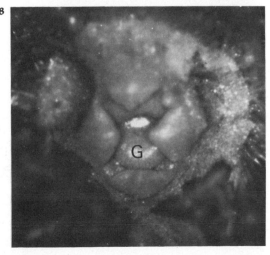

rIv

P
P
P
P

G

2/JUGLANDALES

JUGLANDACEAE

Juglans cinerea L.
(butternut, white walnut)

Female flower

Floral formula:
$\ast$ Iv P_4 $\bar{G}_{(2)}$ O_1

Male flower

Floral formula:
$\cdot | \cdot$ P_6 A_{8-16}

Sequence of primordial inception
Female flower: Iv, $P_{girdling}$, P_{1-4}, G_{1-2}, O_1
Male flower: see text

DESCRIPTION OF FLORAL ORGANOGENESIS

Female flower

The floral apex is initiated in the axil of a bract located close to the inflorescence apex (1). It flattens and two crescent-shaped involucral primordia, each with two distally situated humps, are formed on the transversal flanks (2, 3). These two primordia are not connected to the margin of the subtending bract (2, 4, 5), nor, initially, are they connected to each other at the adaxial flank of the floral apex (2). Subsequently, growth occurs between and adaxial to the two involucral primordia producing either a single protuberance (4) or a two-lobed hump (5). This results in the formation of a single distally-lobed u-shaped primordial involucre. Much later both the subtending floral bract and the lobed involucre are carried up on the ovary wall.

Between the primordial involucre and the subtending bract the floral apex is, at first, flat (3, 5). It becomes concave (4) and then a more-or-less even ridge is formed inside the involucre and bract (6). Four oppositely positioned perianth members form simultaneously on the ridge. The perianth members remain separate from each other, elongate and incurve (7, 8). Two gynoecial primordia form opposite each other in the median plane of the concave floral apex (7). They elongate forming the stigmas and growth occurs between the margins at their base (8, 9). Upward growth occurs in a zone

---

Female flower

1 Top view of a young floral apex. A portion of the bract has been removed. x 246

2 Top view of a young flower bud. The bract has been removed to reveal the flat floral apex flanked on two sides by two bracteolar primordia. The arrowhead points to the region which, in later stages, is occupied by a primordium. x 146

3-5 Later stages of involucre development. The bract has been removed in figures 4 and 5. Note the concavity of the floral apex in figure 4. The arrows in figure 4 point to the gap between the bract and bracteoles. Figures 3-4, x 146. Figure 5, x 85.

6 Top view of floral bud with involucre removed showing perianth inception. x 246

7-8 Later stages of perianth development and gynoecial inception (marked by black arrowhead in figure 7). Figure 7, x 246. Figure 8, x 146.

9 The involucre and two perianth members were removed showing a later stage of gynoecial development. x 146

10 Side view (transversal) of the primordial ovule. The ovary wall and a portion of the primary septum have been removed. x 146

12 Top view of the young ovule showing integument inception. The large arrow in figure 11 points to the region in which the packing tissue forms. Figure 11, x 85. Figure 12, x 146.

14 Side view of the ovule. In figure 13 a portion of the ovary wall remains. The packing tissue (marked by black arrowheads in figure 14) is clearly evident. Figure 13, x 85. Figure 14, x 146.

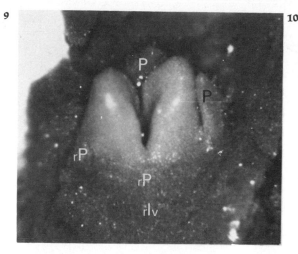

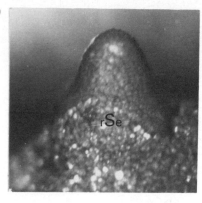

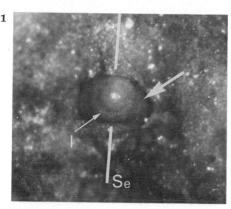

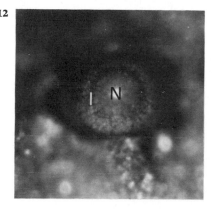

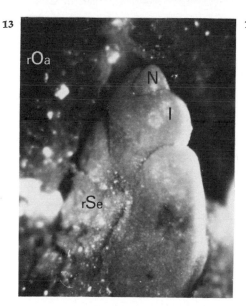

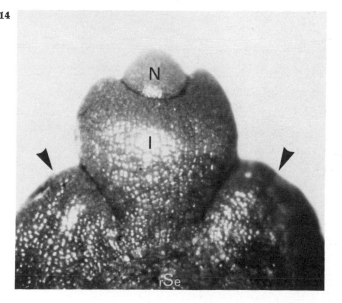

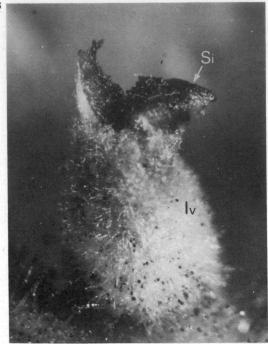

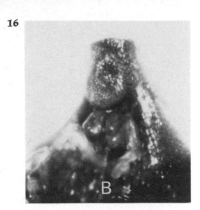

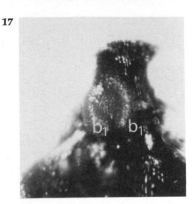

situated beneath the perianth and involucre so that the wall elongates as a cylinder carrying at its tip the perianth and involucre. The ovary then is inferior (9, 15). The inner margins of the gynoecium at the tip of the ovary become appressed and close the locule.

At the base of the ovary the floral apex becomes transformed into an ovular primordium. The primordial ovule is somewhat longer in the median plane than in the transversal plane (10). Before integument formation a narrow septum forms in the transversal plane due to an upgrowth at the base of the young ovary (10, 11). As the ovule grows upward the septum follows resulting in two chambers abaxial and adaxial to the ovule in the median plane. These cavities become filled with tissue formed from growth on the median flanks of the funiculus (11, 13). This growth increases outward in the median plane and upward as the ovule matures at the time of fertilization (13, 14). A single integument arises as a circular bulge just below the ovular apex (11, 12). Further growth results in a bilobed integument.

Male flower
The bracts are arranged in a tight spiral on the inflorescence axis. A floral apex is formed at the base of each bract. A prominent curved asymmetrical bracteolar ridge first appears on one side of the floral apex (16). Following this a similar curved bracteolar ridge forms opposite the first primordial bracteole (17). A third bracteole (tepal) primordium, b_2, arises on the adaxial side of the floral primordium alternate to the two previously formed bracteoles (18). Opposite this primordial bracteole, a fourth bracteole, b_3, may be initiated (19) or may not be formed (20). This bracteole varies greatly in size (22, 24, 25, 29). Subsequently, a pair of oppositely placed bracteole primordia, b_4, are initiated simultaneously on the flanks of the floral apex (36). These

bracteoles develop slowly and do not attain a large size (37, 38, 40).

Just as the number of stamens varies, so does the pattern of stamen initiation. An outer whorl of up to five pairs forms before the inner stamens are initiated (31). The stamens of the outer whorl do not form simultaneously. Relative to the bracteoles and the floral apex, the first stamen-pair forms simultaneously on either side and adaxial to b_2 (19, 20). The second stamen pair forms adaxial to b_1 and to one side of the bracteoles (33, 34). Stamen pair three is initiated adaxial and on either side of b_3, that is, beside stamen pair two (21). Stamen pairs four and five arise possibly simultaneously adaxial and on either side of b_4 (23, 25, 26). One of these two pairs may not form (24, 27). Stamen six arises opposite b_2 and centripetally to the first stamen pair (27, 28, 29).

Stamens formed centripetally to the outer whorl vary greatly in number, arrangement and sequence of initiation. Irregular activity at the floral apex occurs just prior to stamen six inception (26-31). The total number of stamens formed after stamen six may vary from two to five (32-35).

Each stamen consists of two thecae and each theca is composed of two pollen sacs (37, 39).

OTHER AUTHORS

There have been no developmental studies of *Juglans cinerea*. *J. regia* has been studied extensively although hardly from a developmental standpoint. No disagreement exists regarding the constituents of the involucre in the female flower. Hjelmquist (1948) mentions the rare occurrence of three bracteoles which is also recorded here. Manning (1948) and Leroy (1955) indicate that sepal aestivation occurs in two sets, two inner transverse sepals and two outer median sepals. This was not observed in my material nor was it observed by Shuhart (1927) for *Carya* or by Nicoloff (1904-05) for *J. regia*. Nicoloff states that the sepals are whorled.

Parmentier (1911) concludes that the Juglandaceae exhibit a single carpel. This is not borne out by this investigation. He further mentions that *J. cinerea* does not have a secondary septum; this is noted also in this study. Herzfeld (1913) reports a transversal position of the stigmas, whereas a median position was observed in this material. Boesewinkel and Bouman (1967), working with *J. regia*, and Shuhart (1932) working with *Hicoria pecan* record that the integument forms as two primordia. Hagerup (1934) contends that the integument first arises on one side of the nucellus. Benson and Welsford (1909) suggest that *J. regia* has two integuments but this was not figured.

Except for Eichler (1878) who mentions the occurrence of only two or three sterile appendages on the male flower there is no disagreement regarding the number of sterile floral parts. However, there is disagreement regarding their interpretation.

.5 Flower at the time of pollination. x 15

Male flower

6 Top view of a portion of the inflorescence bract with a bracteole primordium on the flank of the floral bud. x 146

7 Top view of a floral bud with two bracteoles.

8 Three floral appendages. x 146

9 b_2 has been removed showing stamen pair 1. A fourth floral appendage, b_3, is shown and a total of two stamen pairs is evident. x 146

0 Floral appendage b_3 is absent in this floral bud. x 146

6 A series of stages showing the inception of the outer whorl of five stamen pairs. The white arrowhead in figure 24 points to a primordium resembling a primordial stamen but occupying the position of floral appendage b_3. Only one member of the floral appendage-pair b_4 can be seen between stamen pairs 4 and 5 in figures 23 and 26. x 146

9 A series of stages showing the formation of stamen six. x 146

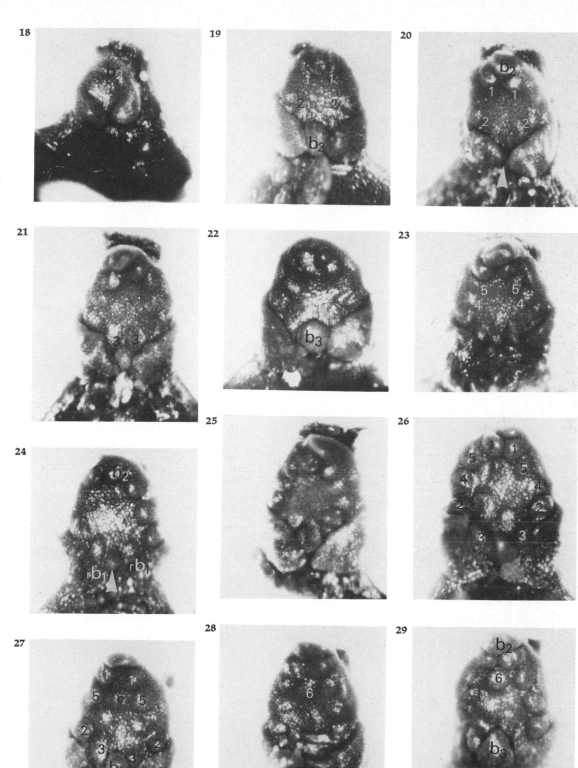

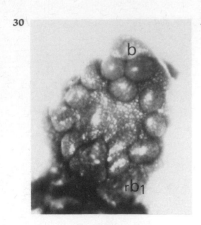

30

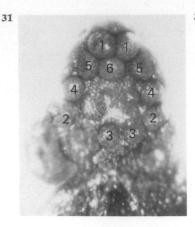

31

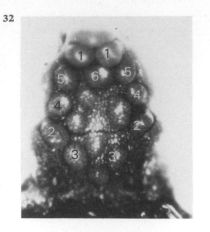

32

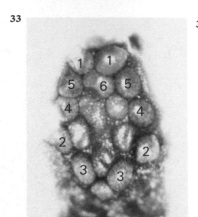

33

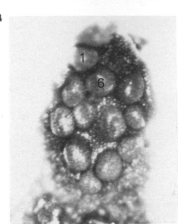

34

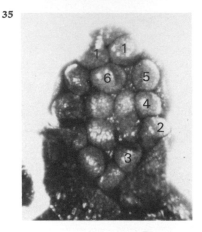

35

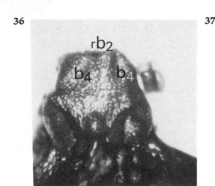

36

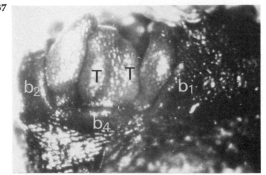

37

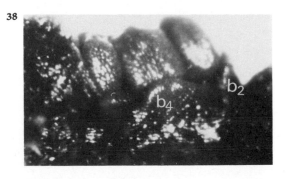

38

No detailed previous work has been done on stamen arrangement or inception. Herzfeld describes two whorls of stamens in *Juglans regia*: an outer one consisting of six pairs of stamens, each pair opposite a tepal, and an inner whorl of three stamens.

Scepotjev (1954) reports the occurrence of bisexual flowers.

Verhoog (1968) describes the development of the gynoecium in the related genus *Engelhardia*.

BIBLIOGRAPHY

Benson, Margaret, and Welsford, E.J. 1909. The morphology of the ovule and female flower of *Juglans regia* and of a few allied genera. Ann. Bot. *23*: 625-633.

Boesewinkel, F.D. and Bouman, F. 1967. Integument initiation in *Juglans* and *Pterocarya*. Acta Bot. Néerl. *16*: 86-101.

Braun, A. 1872. Über den inneren Bau der Frücht der Juglandeen. Bot. Zeit. *30*: 371-375.

Candolle, C. De. 1862. Mémoire sur la famille des Juglandées. Ann. Sci. Nat., Bot. 4ᵉ Sér., *18*: 5-48.

Eichler, A.W. 1878. *Blüthendiagramme*. Leipzig.

Engler, A. and Prantl, K. 1894. *Die natürlichen Pflanzenfamilien*. Leipzig.

Hagerup, O. 1934. Zur Abstammung einiger Angiospermen durch Gnetales und Coniferae. Biol. Medd. udg. af Danske Vidensk. Selsk. *11*: 4, Köbenhavn.

Herzfeld, S. 1913. Studien über Juglandaceen u. Julianaceen. Denkschr. Akad. Wiss. Wien *90*: 301-317.

Hjelmquist, H. 1948. Studies on the floral morphology and phylogeny of the Ameńtiferae. Bot. Notiser, Suppl. *2*: 1-171.

Janchen, E. 1950. Die Herkunft der Angiospermen – Blüte und die systematische Stellung der Apetalen. Oesterr. Botan. Zeitschrift *97*: 129-167.

Karsten, G. 1902. Über die Entwickelung

der weiblichen Blüthen bei einigen Juglandaceen. Flora *90*: 316-333.

Langdon, La Dema M. 1939. Ontogenetic and anatomical studies of the flower and fruit of the Fagaceae and Juglandaceae. Bot. Gaz. *101*: 301-327.

Leroy, S.F. 1955. Etude sur les Juglandaceae. Mém. Mus. Nat. Hist. Nat., Ser. B, Bot. *6*: 1-246.

Manning, W.E. 1940. The morphology of the flowers of the Juglandaceae. II. The pistillate flowers and fruit. Amer. J. Bot. *27*: 839-852.

– 1948. The morphology of the flowers of the Juglandaceae. III. The staminate flowers. Amer. J. Bot. *35*: 606-621.

Nagel, K. 1914. Studien über die Familie der Juglandaceen. Bot. Jahrb. *50*: 459-530.

Nast, C.G. 1935. Morphological development of the fruit of *Juglans regia*. Hilgardia *9*: 345-381.

Navashin, S. 1897. Über die Befruchtung bei Juglans. Trav. Soc. Imp. Sci. Nat. Pétersbourg, *28*: 1.

Nicoloff, M. 1904-05. Sur le type floral et le développement du fruit des Juglandées. Jour. Bot. Paris, *18* (1904): 134-152, 280-285; *19* (1905): 63-68, 69-84.

Parmentier, P. 1911. Recherches anatomiques et taxinomiques sur les Juglandacées. Rev. Gén. Bot. *23*: 341-364.

Scepotjev, F.L. 1954. Bisexual flowers of walnut. Priroda *43*: 92-94 (In Russian).

Shuhart, D.V. 1927. The morphological differentiation of the pistillate flowers of the pecan. Jour. Agric. Res. *31*: 687-696.

– 1932. Morphology and anatomy of the fruit of *Hicoria pecan*. Bot. Gaz. *93*: 1-20.

Van Tieghem, P. 1869. Anatomie de la fleur femelle et du fruit du noyer. Bull. Soc. Bot. France, *16*: 412-419.

Verhoog, H. 1968. A contribution towards the developmental gynoecium morphology of *Engelhardia spicata* Lechen. ex Blume (Juglandaceae). Acta Bot. Neerl. *17*: 137-150.

39

40

31 Two stages after stamen six formation showing pronounced irregular activity at the floral apex. x 146

35 Stamen formation within the outer stamen whorl. In figures 18 and 21 four stamens occur; in figures 19 and 20 are seen five and two stamens respectively. x 146

36 Floral appendage b_2 has been removed to show floral appendage pair b_4. x 146

38 Side views of older floral buds showing formation of thecae (T). x 146

39 Top view of the flower just prior to anthesis. x 20

40 Flower after anthesis showing the floral appendages. x 15

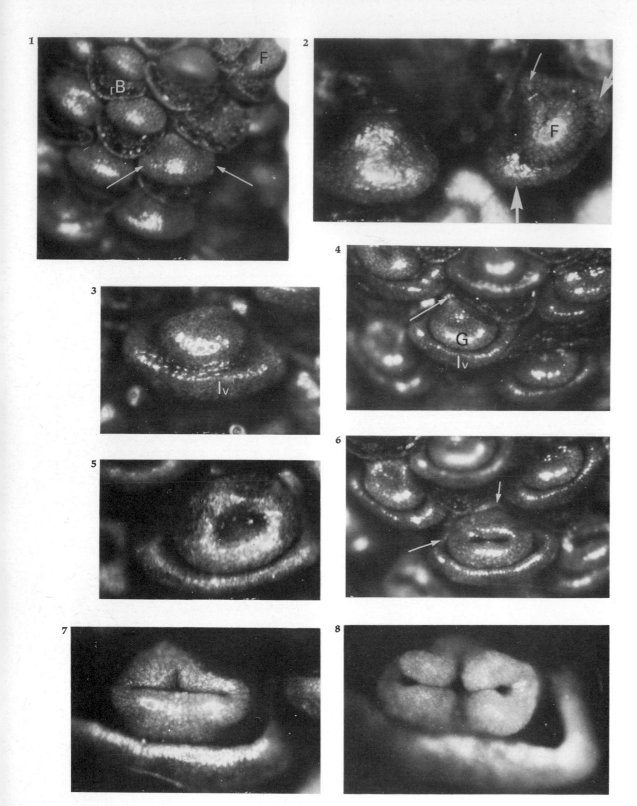

3/SALICALES

SALICACEAE

Populus tremuloides Michx.
(quaking aspen)

Female flower

Floral formula:
$+$ Iv $\underline{G}(2)$

Male flower
Note: The number and arrangement of stamens varies.

Floral formula:
$\cdot | \cdot$ Iv A7-11

Sequence of primordial inception
Female flower: Iv, G_{girdling}, G1-2, O1-∞
Male flower: Iv, A variable, see text

DESCRIPTION OF FLORAL ORGANOGENESIS

Female flower
An ovoid floral apex is initiated in the axil of a bract (1) and enlarges in the transversal plane on its abaxial flank. This results in a three cornered apex with a flat upper surface and a rounded abaxial flank (1). The center of this apex increases in height and becomes rounded (2). This results in a three cornered involucre surrounding the floral apex. The involucre increases in size and depth as the result of growth occurring first between the three corners (3, 4) and then evenly within the involucre (4, 5, 6, 7). Subsequently the involucre becomes asymmetrical, that is, winged, on its abaxial side while the adaxial portion becomes closely appressed to the ovary (9).

The floral apex, meanwhile, continues to increase in height and flattens out (3). A girdling gynoecial primordium then forms about the floral apex (4, 5, 6). While this primordium increases in height, growth within it shifts toward the inner margins at the median plane (6) thus forming two septa. The two semi-circular stigmas and styles are formed on either side (transversally) of these septa (7, 8, 10). The inner margins of the stigmas become appressed as do the two median septa (13, 14) and the ovary wall is formed as a cylinder beneath the styles (9).

The septa are not apparent at the base of the locule (11) but appear higher up on the ovary wall (12, 13, 14). The ovular primor-

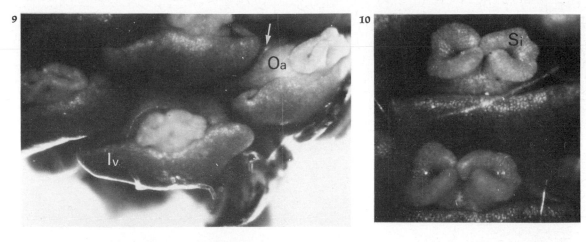

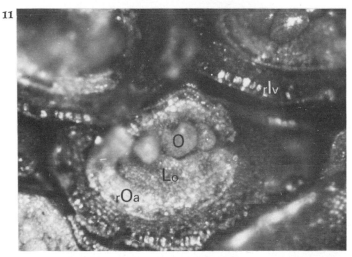

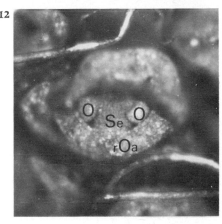

Female flower

1 Side view of the distal region of an inflorescence with bracts removed showing floral buds at various stages of development. The arrows indicate two of the corners of the primordial involucre. x 146

2 Top view of two floral buds showing stages in involucre development. The large arrows indicate the transversal corners of the involucre and the small arrow indicates the posterior involucral bulge. Note the pronounced domed floral apex. x 246

3 Front view of the floral bud. Note the flat floral apex. x 246

4 Top view showing the girdling gynoecial primordium. The arrow points to the posterior portion of the involucre. x 146

5 Top view of the gynoecial primordium. x 246

6-8 Top views of floral buds showing successive stages in the development of the stigmas and gynoecium. The arrows point to the involucre. Figure 6, x 146. Figures 7-8, x 246.

9 Top and side views of floral buds showing the involucre and ovary wall. x 85

10 Top view of two flowers showing the stigmas. x 146

11 Top view of a flower with the ovary wall removed revealing the base of the ovarian chamber. The primordial ovules are attached to the base of the ovary wall. x 146

14 A portion of the ovary wall has been removed showing successively more distal portions of the ovary. In figure 12 the primordial ovule is attached to the side of the septum. In figures 13-14 the septa from each half of the ovary are appressed. x 146

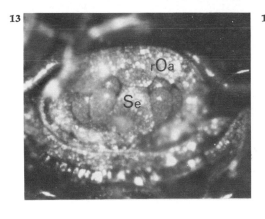

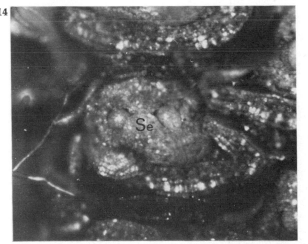

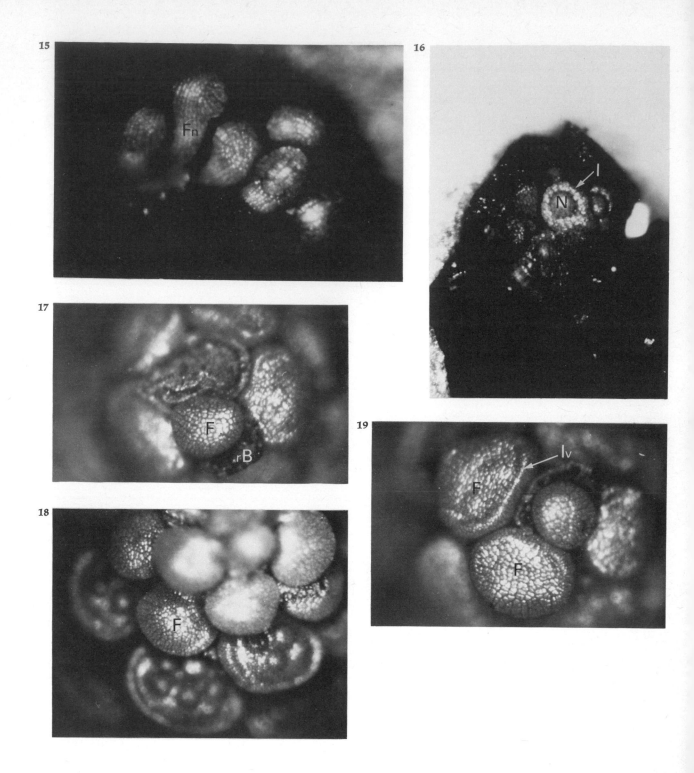

dia arise at the base of the locule in the septal plane (11) as well as on both sides of the septa higher up on the ovary wall (12). The single integument is initiated as a semicircular structure about the tip of the ovular primordium (16) which terminates a long funiculus (15).

Male flower
The floral apex is initiated as a somewhat triangular-shaped body in the axil of a bract (17, 18). Growth occurs rapidly at the flank of the floral apex – more rapidly in the transversal plane than in the median plane (18, 19, 20). In some cases the initiation of the involucre may be seen as a ridge along the adaxial flank of the flat floral apex (19, 20), or in other cases as two ridges on the transversal flanks (18, 21). Ultimately the involucral primordium girdles the floral apex (22, 23).

Stamen initiation occurs bidirectionally; however, only one to three stamens may form adaxial to the first formed primordial stamen. The number of stamens formed and the positional relationships of the stamens are variable. First a single stamen may arise (22, 23, 25), or a pair of stamens may form (24, 28). In the former case this is

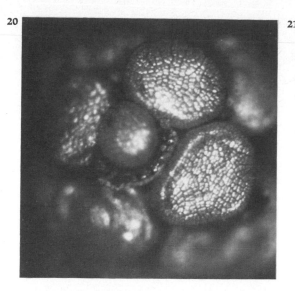
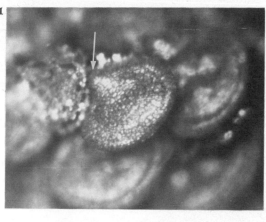
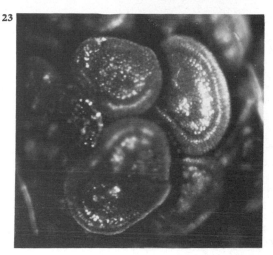
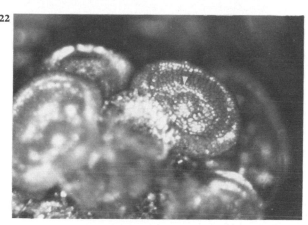
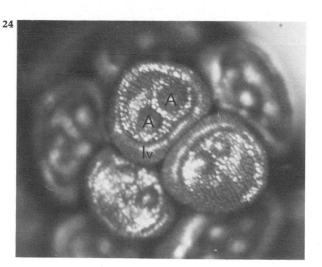

15 The ovary has been removed showing the side view of the ovules. x 146
16 Top view of a young ovule with a primordial integument. x 146

Male flower
18 Top views of an inflorescence with bracts removed showing the young floral bud. x 146
21 Top views of young floral buds showing stages in the development of the involucre. The arrow in figure 21 points to the posterior portion of the involucre. x 146
24 Early stages of stamen inception. An anterior stamen ridge is present in most floral buds. Generally two or three stamen primordia are visible posterior to the stamen ridge. x 146

followed by a pair of stamen primordia forming on both sides of the first primordial stamen, either adaxial to it (22, 28, 29) or abaxial to it (25). This is followed by, or occurs concurrently with, the establishment of a slightly lobed ridge on the abaxial side of the first formed stamen(s) (22-25). Up to five stamen primordia may arise on this ridge more or less simultaneously (25, 27, 29, 30, 31, 32). Some variation occurs in which stamen primordia are formed on this ridge in a less regular spatial arrangement (29, 33, 34) in which case stamens arise abaxial to the previously formed primordial stamens.

The pollen sacs are formed on either side of the stamen primordia and the stamens are extrorse (34, 35). Note that the pollen sacs of the posterior stamen are directed toward the inflorescence axis, that is, they are oppositely directed compared to the other stamens (34).

OTHER AUTHORS

Few detailed developmental studies of the reproductive structures of this genus have been made. Previous studies of the development of the female flower agree with my observations. Graf (1921) found flowers with three and four carpels. He also describes a rudimentary inner integument which is formed immediately after the outer one in *P. canadensis* and *P. canescens*. This was not recorded for *P. alba* or *P. tremula*.

Results based on median longitudinal sections of the male flower obtained by Hegelmaier (1880), Graf (1921), and Nagaraj (1952) indicate that stamen inception proceeds in two directions from the center of the floral bud, i.e. centrifugally. My findings indicate that stamen inception may occur bidirectionally but most of the stamens are initiated in rows towards the abaxial side of the floral bud.

Aubert (1875) and Hagerup (1938) describe the floral development of *Salix*.

BIBLIOGRAPHY

Aubert, P.H. 1875. Organogénie de la fleur dans le genre *Salix*. Adansonia 11: 183.

Eichler, A.W. 1878. *Blüthendiagramme*. Leipzig: Wilhelm Engelmann.

Fisher, M.J. 1928. The morphology and anatomy of the flowers of the Salicaceae. Amer. J. Bot. 15: 372-394.

Graf, J. 1921. Beiträge zur Kenntnis der Gattang *Populus*. Beih. Bot. Centralbl. 38: 405-454.

Hagerup, O. 1938. On the origin of some Angiosperms through the Gnetales and Coniferae. III. The gynaecium of *Salix cinerea*. Kgl. Danske Vid. Selsk. Biol. Medd. 14: 1-34.

Hegelmaier, Fr. 1880. *Ueber Blüthenentwicklung bei den Salicinee*. Württemberg. naturwiss. Jahresheft. p. 205.

Hjelmquist, H. 1948. Studies on the floral morphology and phylogeny of the Amentiferae. Bot. Notiser, Suppl. 2: 1-171.

Nagaraj, M. 1952. Floral morphology of *Populus deltoides* and *P. tremuloides*. Bot. Gaz. 114: 222-243.

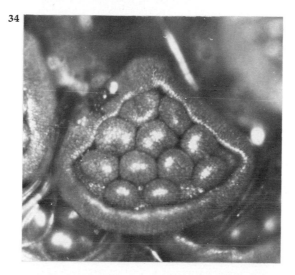

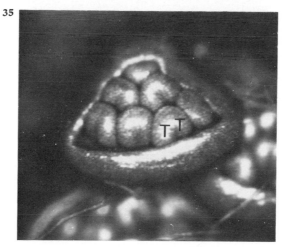

33 Early stamen development. The arrow points to the anterior stamen ridge. x 146

34 Stages in stamen development. Note stamen inception occurs progressively towards the anterior portion of the flower. x 146

35 Top view showing extrorse stamens with thecae. x 146

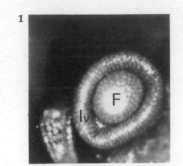

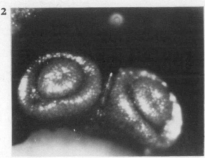

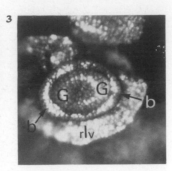

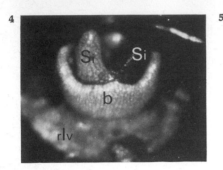

4 / FAGALES

BETULACEAE

Ostrya virginiana (Mill.) K. Koch
(hornbeam)

Female cymule

Floral formula:
$+ \text{Iv} \, P_4 \, \overline{G}(2) \, O_2$
Note: P4 are referred to
as bracteoles.

Male cymule

Floral formula:
$+ \text{ or } \cdot | \cdot \, A_{3\text{-}4}$

Sequence of primordial inception
Female flower: Iv, P_{girdling}, P1-2, 3-4 (variable), G1-2,
O1-2
Male flower: variable, see text

DESCRIPTION OF FLORAL ORGANOGENESIS

Female flower
Two floret apices arise in the axil of a primary bract. The first appendage formed at the apex is a semi-circular primordial involucre (1) which surrounds almost completely the domed floret apex. The open end of each involucre is directed away from the cymule axis; the two florets are mirror images (2). The involucre then forms a cylindrical structure overgrowing and enclosing the floret apex and all subsequently formed appendages.

The floret apex flattens and forms a distally uneven primordium that girdles the apex (2). Two primordial bracteole bulges form at each transversal pole of the girdling primordium (3). Subsequently, two bracteole primordia arise simultaneously in the median plane on the girdling primordium (4). Occasionally more than four bracteoles

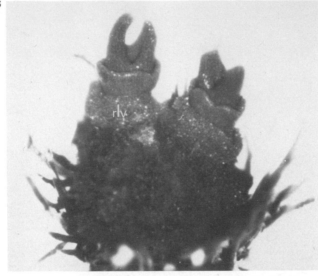

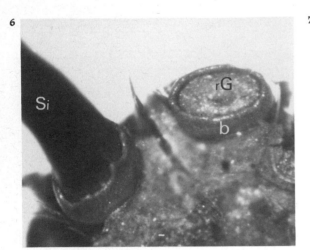

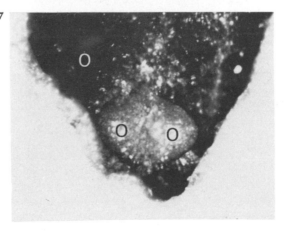

may form (5); however, growth occurs between the bracteoles forming a distally ridged girdling structure (6).

Opposite the first formed bracteole primordia two crescent shaped gynoecial primordia are initiated simultaneously. It is difficult to say if they form after or simultaneously with the first bracteole pair (3). The distal region of the two primordia forms the stigmas. Growth occurs between the gynoecial primordia forming a cylindrical ovary wall which grows upward and encloses the locule when the distal margins of the gynoecium become appressed (5). The bracteoles are carried up with the ovary wall as the latter undergoes extensive

---

Female flower

1 Top view of a floral bud with a dome-shaped floral apex surrounded by a primordial involucre. The bract has been removed. x 146

2 The bract has been removed revealing two floral buds which are mirror images of each other. Note the uneven bracteolar ridge girdling the apex. x 146

3 The bract and involucre have been removed to show two bracteoles and two opposite gynoecial primordia. x 146

4 Side view showing three of four bracteoles and two primordial stigmas. x 146

5-6 Side views of florets with bracts and involucres removed to show stages in the development of stigmas and involucral ridge. In figure 6 one of the gynoecia has been removed revealing the bracteolar ridge. x 85

7-8 The ovary wall has been removed to show successive stages in ovule inception. The arrowhead points to the portion of the septum which has separated from the ovary wall. Figure 7, x 146. Figure 8, x 85.

9 The ovary wall has been removed showing the ovule with the single integument. x 85

10 Side view of the inflorescence at the time of pollination. The stigmas are visible above the subtending bracts. x 20

Male flower

11 Top view of bract and an elongate cymule primordium. x 146

12 Three floret apices are evident. x 146

13 The arrowhead marks the position that usually is occupied by a floret apex. x 146

14 Two floret apices are poorly separated (marked by arrowhead). x 146

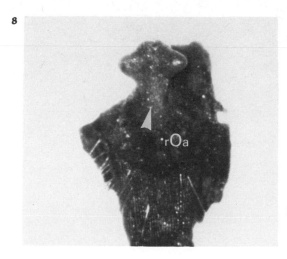

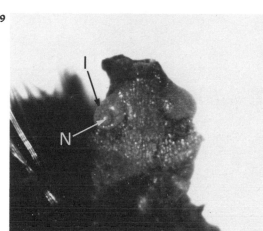

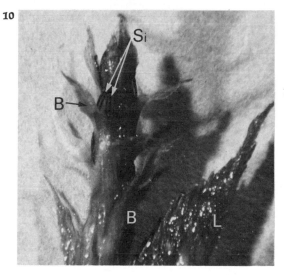

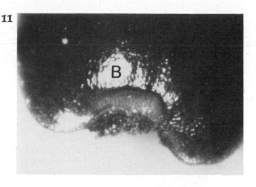

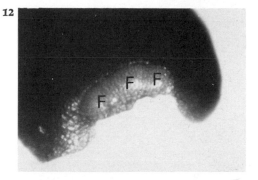

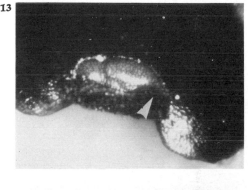

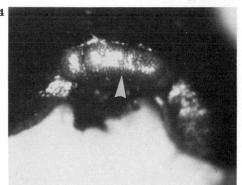

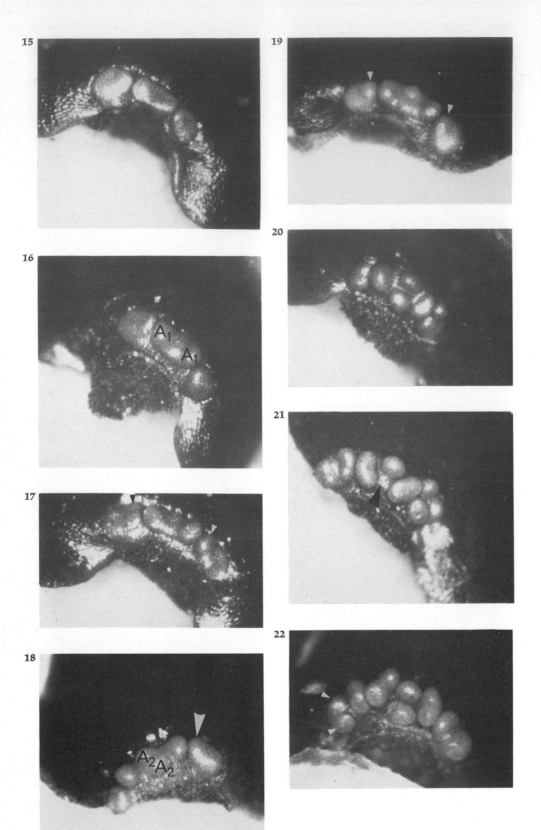

growth; growth has occurred in a zone beneath the gynoecium and bracteoles (4, 5, 6). The involucre is not raised up with the gynoecium and bracteoles.

At the center of the base of the locule the floral apex is transformed into a small primordium which forms two transversally oriented ovular primordia (7). This structure becomes elevated on a stalk (8) in continuity with a septum which in turn is continuous with the ovary wall in the median plane. Each primordial ovule forms a single integument (9). The two ovules, situated at the same level on the common stalk, become directed away from the transversal plane, that is, toward the abaxial side for one ovule and toward the adaxial side for the other ovule (9).

Male flower
The cymule primordium arises on the base of a large subtending bract (11). The primordium expands transversally and forms three floret primordia (12). Some variation occurs and only two floret primordia may form (13, 29), or the separation of one floret from another may be indistinct (14). Subsequent growth appears to be restricted to these regions which tend to act independently from one another.

Each floret apex becomes flat due to expansive growth at the periphery of the apex (15). Two stamen primordia first appear simultaneously occupying the entire transversal flanks of the central floret (15, 16). Subsequently, two oppositely placed median stamen primordia arise simultaneously alternate to the two first formed primordial stamens (18, 19, 20) or the adaxial stamen primordium may form first (17). Sometimes only one primordium occurs in the median position of the central floret and this generally is the abaxial one (21, 24). However, only the adaxial one of the pair may be present (26). Some cases occur in which one of the transversal primordia of the central floret does not form

(26, 27), or in which the positional relationship of the stamen primordia is upset (28).

On the transversal floret apex a stamen primordium first forms on the interior adaxial flank (15-20). Two stamen primordia may arise on this side of the floret apex (23). The next stamen primordia to form arise on the outer transversal flanks of each transversal floret; they form simultaneously and opposite each other (21, 22, 26). In some cases only one stamen may form on the outer transversal flank (23, 27).

Early in its development the stamen primordium bifurcates forming two thecae (25). As the stamen matures it is the filament which appears dichotomously branched; each branch terminates in a theca (30). Each theca forms two pollen sacs (25).

---

6 Two transversal stamen primordia are evident on the central floret. On each of the transversal florets a stamen primordium is visible on the inside flank. x 146

7 A single stamen primordium is visible between the two transversal primordial stamens on the central floret. The arrowheads mark primordial stamens on the transversal florets. x 146

0 Stages in the development of the anterior-posterior stamen pair on the central floret. Notice in figures 18 and 20 the difference in size of the stamens on the interior flank of the transversal florets. x 146

2 The black arrowhead points to the position usually occupied by the adaxial stamen in figure 21. Note the two stamen primordia on the outer flank of each transversal floret. The transversal stamens on the central floret are at different stages of dichotomy. x 146

5 Different stages of stamen dichotomy, theca formation, and pollen sac formation. x 146.

9 Representatives of unusual stages of floret development in which the region between two florets is not distinct in figure 26; the central floret is missing a transversal stamen primordium in figure 27; the stamens of the central and right floret are aligned in two parallel rows in figure 28; the right floret is absent in figure 29. x 146

• The arrow points to the filament dichotomy of a stamen just prior to anthesis. x 85

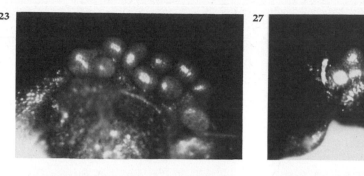

23

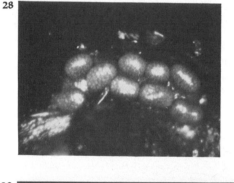

27

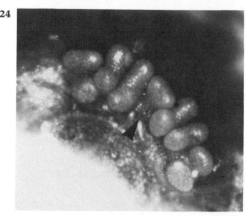

24

28

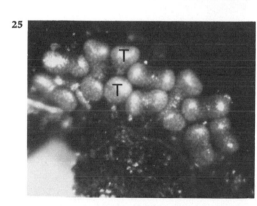

25

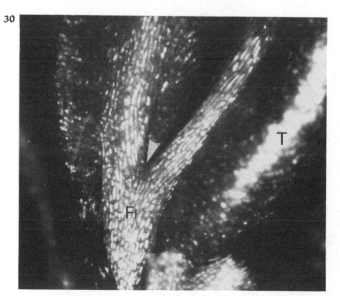

29

30

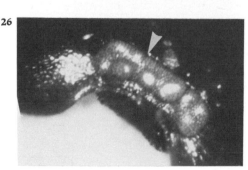

26

OTHER AUTHORS

Endress (1967) examined the later stages of
ovary development in *Ostrya carpinifolia*
including the development of the marginal
placentae. He observes that the two ovules
may form either on one placenta or one
ovule may form on each placenta. This has
also been noted by Baillon (1876-79) for
Betula, Corylus, and *Alnus,* and by Streicher
(1918) and Schacht (1854) for *Betula.*
Wolpert (1910) describes the floral organogenesis of *Alnus alnobetula* and *Betula.*
Hagerup (1942) examined the development
of the ovules and fruit of *Corylus avellana*
and *C. maxima* from sections. He notes that,
although the placenta arises at the base of
the ovary, it becomes parietal and the ovules
are located on opposite sides of the ovary
wall. He further states there is no septum,
that there only appears to be a septum due
to the placentae being appressed. This differs from my material in that the ovules
arise on a column which extends as a continuation of the floral apex and which is
continuous with the ovary wall in the
median plane by a septum.

It is the presence or absence of sterile
appendages, the secondary, tertiary, and
quaternary bracts (tepals, perianth), which
has attracted the greatest attention. Abbe
(1935), in reference to his figure 136, states
that the tertiary bracts are more conspicuous than the secondary bracts. This is
not evident from his figure. Abbe (1938,
figure 88) located the septum of *Ostrya* in
the transversal plane whereas the septum
in my material is median. No evidence of a
secondary bract can be found in my material of the female flowers (see Abbe 1938,
figure 88).

Regarding the male florets, Hjelmquist
(1948) states that, although variations
occur, a perianth (two median and two
transversal) is found in *Ostrya.* Abbe
(1935) has figured vascular bundles which
he interprets as belonging to aborted

perianth members. In my material no sterile
appendages are formed in association with
the florets. The stamen arrangement on the
lateral flowers of Abbe's material (Abbe
1938, figures 89, 90) is not the same in my
material.

BIBLIOGRAPHY

Abbe, E.C. 1935. Studies in the phylogeny
of the Betulaceae. Bot. Gaz. *97*: 1-67.
– 1938. Studies ... Bot. Gaz. *99*: 431-469.
Baillon, H. 1876-79. Traité du développement de la fleur et du fruit. Adansonia
12: 1-19.
Endress, P.K. 1967. Systematische Studie
über die verwandtschaftichen Beziehungen zwischen den Hamamelidaceen
und Betulaceen. Bot. Jb. *87*: 431-525.
Hagerup, O. 1942. The morphology and
biology of the *Corylus*-fruit. Kgl. Danske
Vid. Selsk. Biol. Medd. *17*: 3-32.
Hjelmquist, H. 1948. Studies on the floral
morphology and phylogeny of the
Amentiferae. Bot. Notiser, Suppl. *2*:
1-171.
Schacht, H. 1854. Entwicklungsgeschichte
der Capuliferen- und Betulineen-Blüte.
In: *Beiträge zur Anatomie und Physiologie der Gewächse.* Berlin.
Streicher, M. 1918. Zur Entwicklungs-
geschichte des Fruchtknotens der Birke.
Denkschr. Akad. Wiss. Wien, math.-nat.
Kl. *95*: 355-367.
Wolpert, J. 1910. Vergleichende Anatomie
und Entwicklungsgeschichte von *Alnus
alnobetula* und *Betula.* Flora *100*: 37-59.

5/FAGALES

FAGACEAE

Quercus rubra L. (red oak)

Female flower
Note: The involucre and the subtending bract have been omitted.

Floral formula:
$* \text{Iv } P3 + 3 \, \overline{G}(3) \, O6$

Male flower

Floral formula:
$* P4\text{-}5 \, A4$
Note: Five perianth members are initiated but usually only four attain maturity.

Sequence of primordial inception
Female flower: P1-3,4-6, G1-3, O1-6
Male flower: P1,2,3,4,5, A1,2,3,4
Note: Six stamens may occur.

DESCRIPTION OF FLORAL ORGANOGENESIS

Female flower

The inflorescence contains a number of bracts arranged spirally on the axis. The floral apices form in the axils of each of two of the lower bracts located on either side, in the transversal plane, of the inflorescence (2). Occasionally only one floral apex is formed (1). The floral apex flattens and assumes a three-cornered appearance as the result of growth occurring at its periphery (1), i.e., two small humps appear simultaneously on either side of the adaxial flank (1), preceding the one on the abaxial flank (1) or arising simultaneously with the outer abaxial perianth primordium (2). In a sim-

ilar manner, three inner perianth members are initiated simultaneously between the outer perianth members (1, 2). The inner and outer perianth members are then raised up on a common base (4). As this occurs three gynoecial primordia arise on the broad flat apex opposite the outer perianth members (4). The margins of these three primordia grow towards each other and become appressed (5). These primordia develop into three stigmas, or occasionally, if four primordia are formed, then a four-stigmatic flower will result (6).

Zonal growth occurs in a region beneath the primordial perianth and gynoecium so that as the wall grows upward as a cylinder, the perianth is carried up with the ovary wall (6, 7). The young ovary becomes closed as a result of the appression of the three gynoecial primordia as they are carried up with the extending ovary wall. Concurrently, growth between and at the base of the three gynoecial primordia initiates the septa. The inward growth of the septa extends up along the wall. Thus, three septa are formed which eventually become appressed at their upper inner margins (9, 10). The septa, however, are not joined at their base (10). Hairs develop on the inner surface of the wall (9, 10).

Two placentae form first as slight protrusions along the base and on each side of

Female flower

1 Top view of an inflorescence showing inflorescence apex, floral apex and bracts. Older bracts have been removed. Two perianth primordia are evident on the inside flank of the floral apex. x 85
2 Top view of an inflorescence with two floral apices showing different stages of development. The black arrowheads mark the inner perianth primordia. x 85
3 Top view of a floral bud showing the perianth primordia. x 85
4 Top view of two floral buds. The black arrow indicates the perianth. Three gynoecial primordia are evident. x 85

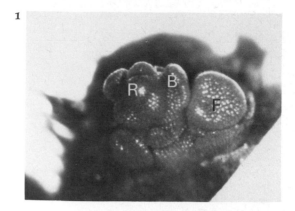

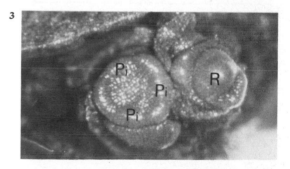

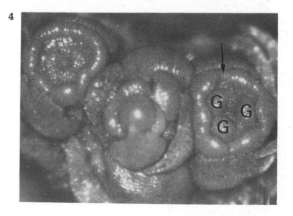

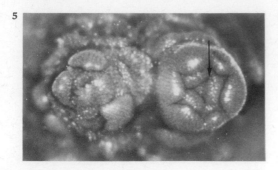

5

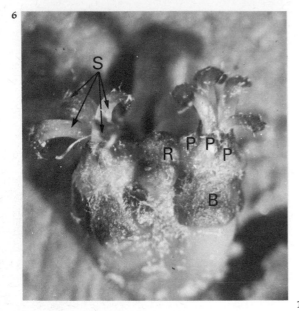

6

S

R P P P

B

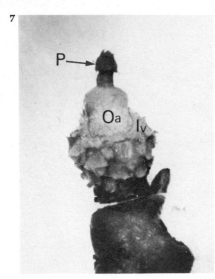

7

P

Oa Iv

8

Iv

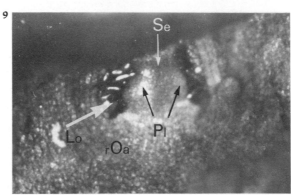

9

Se

Lo rOa Pl

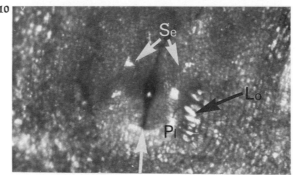

10

Se

Lo

Pl

11

Ii

Io

12

Lo

N

Io Ii

the septa (9, 10). This means that in one locule there are two placentae – one from each septum. The ovule develops as a further extension of the placental bulge (11). At first this young ovular primordium forms as an elongated structure, rounded at its lower region and slightly convoluted on its upper surface. It is difficult to determine the sequence of integument initiation because of the peculiar bulges occurring on the primordial ovular surface (11). The inner integument girdles the nucellus and the outer integument is initiated first as a crescent-shaped structure (12). At the time of fertilization the margins of the outer integument, toward the open end, greatly increase in height thereby forming a lip (13).

The involucre (7, 8), surrounding the gynoecium and perianth, is first formed as a few separate primordia in the axil of the bracts located at the base of the floral axis (14, 15). Growth occurs between these axillary involucral primordia to form a solid ring around the base of the flower (16, 17, 8). Subsequently, small bract-like appendages arise first on the outer flank of the

---

5 Top view of floral bud with three appressed gynoecial primordia. x 85

6 Two flowers each with four stigmas at the time of pollination. x 85

7 Fruits one year after pollination. A portion of the involucre has been removed showing the ovary wall. Note that the scale-like perianth is carried on a small distal extension of the ovary wall. Scales below the involucre have been removed. x 20

8 Top view of two female structures one year after pollination showing the involucre. The perianth and stigmas have become scale-like. x 38

9 A portion of the ovary wall has been removed to show a head-on-view of one septum. A placenta occurs on each side of the septum and extends into different locules. Note the hairs from the inner surface of the ovary wall. x 146

10 A portion of the ovary wall surrounding one locule has been removed revealing the placentae from two septa in face view. The white arrow points to the base of the ovarian cavity. x 146

involucral apex (16) and subsequently, as this structure becomes older, the involucral bracts form centripetally on the upper surface in a phyllotactic sequence consisting of numerous parastichies (18, 19, 20).

Male flower
The ovate-shaped floral apex is initiated in the axil of a small ridge-shaped bract close to the inflorescence apex (21). It enlarges in all planes and renders the subtending bract inconspicuous by comparison. The perianth members arise on the flanks of the floral apex in a spiral sequence. The plastochron is very short, particularly between the initiation of the second, third, and fourth perianth members (22, 25). After the fifth perianth member has formed, growth extends between P_3 and P_5 (24, 25, 27, 28), and even in some cases between P_2, P_3 and P_5 (27), thus forming one large abaxial perianth member. This type of

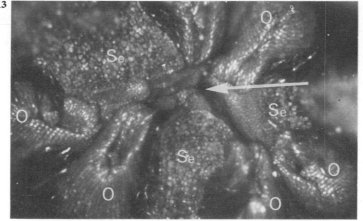

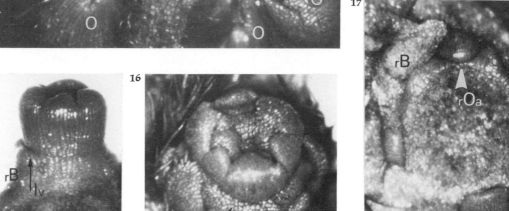

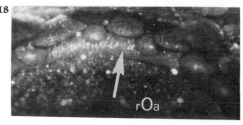

12 Top views of two young ovules at different stages of development. Note the outer integument does not completely envelop the nucellus and inner integument. x 146

13 The ovary wall has been removed to show a top view of the ovules just after fertilization. Note the 'lip' of the outer integument and the three septa. The white arrow indicates the region of the base of the ovary where the septa become appressed. x 85

15 The bracts have been removed showing a side view of two floral buds at different stages of development. The arrow points to the axillary ridge-like structures constituting the primordial involucre. x 85

16 Top view of a floral bud showing the involucral apex (marked by black arrow) and a few involucral scales. x 85

19 The ovary has been removed showing stages in the development of the involucre viewed from the inside. Figures 17 and 18 are top views. Figure 19 is a side view of the inside. x 85

20 Side view of the outside portion of the involucre corresponding in time to the stage pictured in figure 19. Some involucral bracts have been removed. x 85

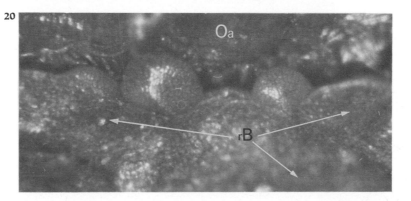

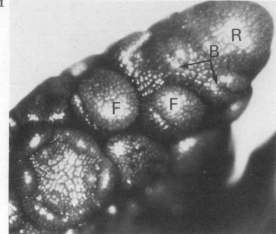

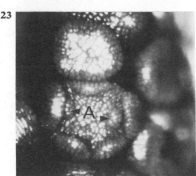

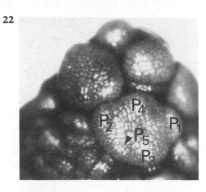

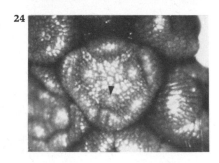

growth ultimately occurs beneath all the perianth members thereby elevating them on a common sheath (30).

The sequence of stamen initiation is quite regular. Four stamen primordia are formed and the last to form is always the adaxial one which is initiated opposite a perianth member (25, 27, 28). The two transversal stamens arise also opposite a corresponding perianth member and probably in a sequence of initiation of the perianth members (25, 27). It is possible that they arise simultaneously (23) or even simultaneously with the single abaxial stamen (21, 24). The abaxial stamen may arise opposite a perianth member (21, 24), or it may form between two perianth members (25-28). Each theca consists of two pollen sacs. The anther is extrorse and subdorsifixed (29).

OTHER AUTHORS

Berridge (1914) shows for *Q. cerris* that the placentae form on the septa and not on a central prolongation of the axis. The latter structure forms after the placentae become noticeable and is continuous with the septa at the base of the ovarian cavity. On the other hand, Langdon (1939; see also Baum and Leinfellner 1953) claims that the details of floral organogenesis in *Q. rubra* and *Fagus americana* are similar. This means that the ovules would be carried at the tip of an axile prolongation and, therefore, are not marginal as claimed by Berridge (1914), Turkel *et al.* (1955) and as found in my material. Both Langdon and Berridge observed an axile stalk or prolongation which is continuous at its base to ingrowths, or margins, of the gynoecial wall, thereby forming the septa. This was not observed in my material of *Q. rubra* which was not examined after the time of integument formation although I have observed it in material of *Fagus grandifolia*. Turkel *et al.* (1955), working with *Quercus alba*, note that the carpels become 'adnate'

to the perianth. They describe the ovules as forming on the 'axile placental folds of the inner carpellary walls' but make no mention of a basal upgrowth or axile prolongation. They noted that the two integuments formed simultaneously. Baillon (1876-79) reports that the five or six members of the perianth of *Q. robur* and *Q. hybrida* form in succession.

The ontogeny of the involucre or cupule of *Quercus* was examined by Schacht (1854) and Baillon (1876-79). The latter found the cupule formed from a single ring primordium. Schwartz (1936) shows the formation of the cupule scales which he had apparently mistaken for primordial flowers. Brett (1964) describes the early formation of the inflorescences, cupule, and in some cases, but not in detail, the floral parts of *Fagus sylvatica* and *Castanea sativa*.

In regard to the male flowers of *Quercus*, Hjelmquist (1948) states that there appears to be no definite positional relationship between the stamens and the perianth members. It should be pointed out that his observations were taken from mature structures only. His observations are not confirmed by my material. Turkel *et al.* (1955) mention that for *Q. alba* the 'sepals arise as a common structure' but make no mention of the sequence or positional relationships of the stamens.

Garrison (1956) made a study of the ontogeny of *Fagus grandifolia*.

BIBLIOGRAPHY

Baillon, H. 1876-79. Traité du développement de la fleur et du fruit. Adansonia 12:1-19.

Baum, H. and Leinfellner, W. 1953. Bemerkungen zur Morphologie des Gynözeums der Amentiferen im Hinblick auf Phyllo- und Stachyosporie. Oesterr. Bot. Zeitschrift 100: 276-291.

Berridge, E.M. 1914. The structure of the flower of the Fagaceae and its bearing on the affinities of the group. Ann. Bot. *28*: 509-526.

Brett, D.W. 1964. The inflorescence of *Fagus* and *Castanea* and the evolution of the cupules of the Fagaceae. New Phytol. *63*: 96-118.

Garrison, H.J. 1956. Floral morphology and ontogeny of *Fagus grandifolia* Ehrh. Ph.D. Thesis. Pennsylvania State Univ.

Hjelmquist, H. 1948. Studies on the floral morphology and phylogeny of the Amentiferae. Bot. Notiser, Suppl. *2*: 1-171.

Langdon, LaDema M. 1939. Ontogenetic and anatomical studies of the flower and fruit of the Fagaceae and Juglandaceae. Bot. Gaz. *101*: 301-327.

Schacht, H. 1854. *Beiträge zur Anatomie und Physiologie der Gewächse*. Berlin.

Schwartz, O. 1936. Entwurf zu einem natürlichen System der Cupuliferen und die Gattung *Quercus L.* Not. bot. Gart, Berlin-Dahlem, *13*: 1.

Turkel, H.S., Rebuck, A.C., and Grove, A.R. 1955. Floral morphology of white oak (*Q. alba*). Pa. Agr. Expt. Sta. Bull. 593, p. 14.

Male flower

21 Side view of the distal region of an inflorescence showing floral buds at different stages of development. Note the rudimentary bracts subtending the floral buds. x 146

22 Top view of a floral bud showing the sequence of perianth inception. x 146

26 Top views of floral buds showing stages in stamen inception. Note the difficulty of determining the number of perianth members on the lower (anterior) portion of the floral buds in figures 24-26. The black arrowhead points to a stamen positioned opposite a perianth member in figure 24 and alternate to two perianth members in figure 26. x 146

28 Stages in the development of the anterior portion of the perianth (marked by arrows). x 146

29 Adaxial view of a stamen just prior to anthesis. x 85

30 A portion of the perianth has been removed to reveal the male flower just prior to anthesis. x 80

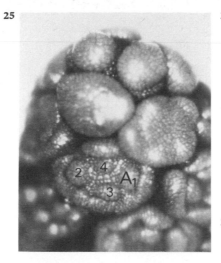

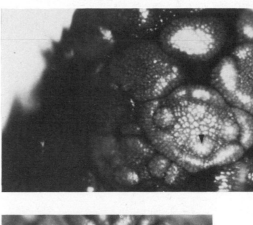

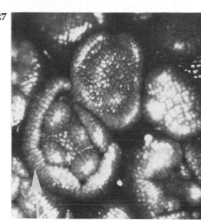

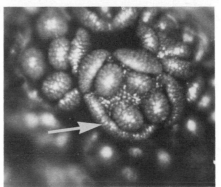

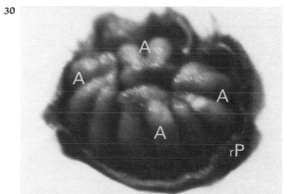

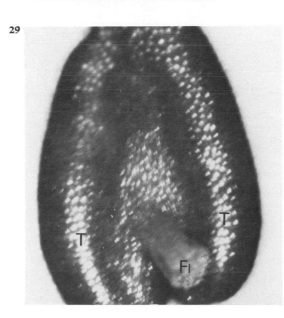

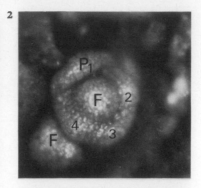

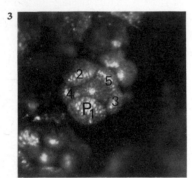

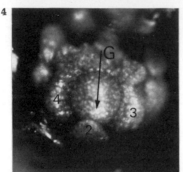

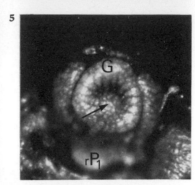

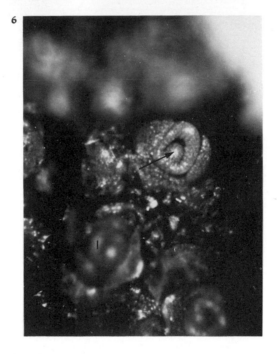

6/URTICALES

URTICACEAE

Laportea canadensis (L.) Wedd. (wood-nettle)

Female flower

Floral formula:
·|· P4 G̲1 O1

Male flower

Floral formula:
＊ P5 A5

Sequence of primordial inception
Female flower: P1,2,3,4, G1, O1
Note: A fifth perianth member may occur.
Male flower: P1,2,3,4,5, A1,2,3,4,5
Note: In some cases only four perianth members and four stamens occur.

DESCRIPTION OF FLORAL ORGANOGENESIS

Female flower
The young floral apex is broadly convex (1). Up to five perianth members are formed in a spiral sequence (3). The plastochron is greater between the initiation of the first and second members than between the formation of the second, third and fourth members (1, 2). Five perianth members do not form in all cases (2, 4) and possibly when five arise then two 'fuse' as the result of growth between them thereby forming four almost oppositely-positioned perianth members (4, 5).

After perianth inception the floral apex becomes more prominently dome-shaped (2). Subsequently it flattens out as the result of growth occurring at the periphery. A bulge then forms at the flank opposite the second-formed perianth member primordium (4). From this gynoecial primordium growth radiates around the periphery of the floral apex forming first a crescent-shaped structure (5), and then a circular structure (6). Subsequently, a cylindrical structure, the ovary, is formed due to upward growth occurring in the lower portion of the gynoecium. The style and stigma originate by postgenital fusion of the margins of the gynoecial primordium (9, 10).

The floral apex resumes growth and becomes transformed into the ovule primordium (5, 6). Two circular integuments are formed from this primordium (8). A variation may occur in which the outer integument does not develop as a uniformly circular primordium and the inner integument develops as a thick cup-shaped structure which overgrows the nucellus (7).

Male flower
The primordia of the five perianth members are initiated in a spiral sequence (11, 12). However, the plastochron is very short.

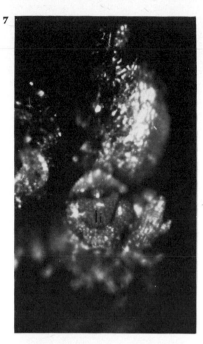

7

9

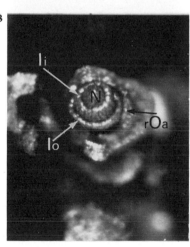

8

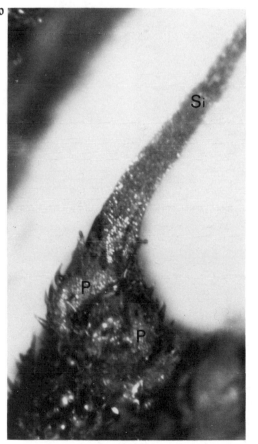

10

Female flower

1-3 Top views of floral buds showing stages in the inception of perianth members. Figures 1 and 3, x 146. Figure 2, x 246

4 Top view of a floral bud showing gynoecium inception. The first-formed perianth member has been removed. x 246

5-6 Top views of a young gynoecium. The arrow indicates the primordial ovule. Figure 5, x 246. Figure 6, x 146

7 An unusual integument form; part of the outer integument has been removed but it is clearly not uniformly circular. The ovary wall has been removed. x 146

8 Top view of an ovule showing two integuments. x 146

9 A young flower with one perianth member removed to show the suture (marked by arrow) and ovary wall. x 85

10 Side view of a mature flower. x 85

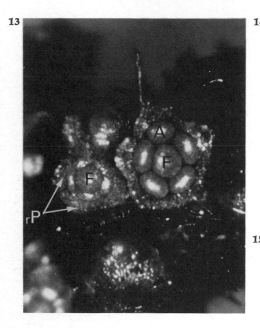

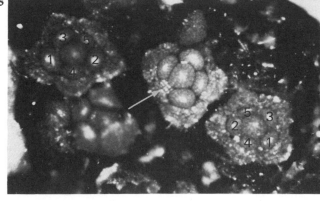

Throughout the initiation of the perianth the floral apex retains its identity and as the perianth enlarges the convexity increases.

The stamens are formed in the axils of the primordial perianth members and the sequence of initiation follows that of the corresponding subtending perianth members. As in the perianth, the sequence may be clockwise (15) or counter-clockwise (14, 15) and the plastochron is very small (13). Sometimes only four stamens are initiated in which case it is possible that only four perianth members have been initiated (15). The stamens are extrorse (16, 17) and the filament is evident along the abaxial surface of the anther (16).

The floral apex continues to enlarge throughout stamen initiation and flattens as the result of growth on its periphery (13). This growth is more prolonged at the periphery than at the center so that at anthesis the apex is surrounded by a ring structure (17).

OTHER AUTHORS

I could find no developmental study of *Laportea* in the literature. Bernbeck (1932) examined the development of the inflorescence of *Laportea* but not the individual flower. Modilewsky (1908) studied the embryology of *Laportea moroides*. In his figure 40 (p. 444) he shows a longitudinal section through the gynoecium with the ovule whose inner integument forms a long tube. Bechtel (1921) examined the anatomy of *Laportea canadensis*.

BIBLIOGRAPHY

Bechtel, A.R. 1921. The floral anatomy of the Urticales. Amer. J. Bot. *8*: 386-410.
Bernbeck, F. 1932. Vergleichende Morphologie der Urticaceen- und Moraceen-Infloreszenzen. Botanische Abhandlungen *19*: 1-100.

Modilewsky, J. 1908. Zur Samenentwicklung einiger Urticifloren. Flora *98*: 423-470.

Rivières, R. 1956. Fleurs et inflorescence de quelques Urticacées. Naturalia Monspeliensia ser. Bot. *8*: 189-204.

— 1958. Les inflorescences chez les Urticacées. Naturalia Monspeliansia ser. Bot. *9*: 163-172.

Male flower

12 Top view of floral buds showing stages of perianth inception. x 146

15 Top view of flowers wih perianth removed showing different stages of stamen development. The arrow in figure 14 indicates the absence of a primordial stamen. x 146

16 Side view of the abaxial surface of a mature stamen (Fl = filament). x 85

17 Top view of a mature male flower. The perianth has been removed. x 85

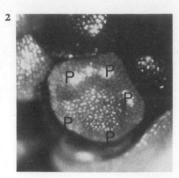

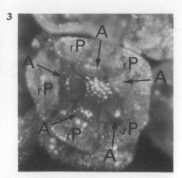

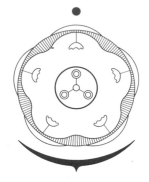

7/SANTALALES

SANTALACEAE

Comandra umbellata (L.) Nutt.
subsp. **umbellata**
(= **Comandra richardsiana** Fern.)
(Richardson's bastard-toadflax)

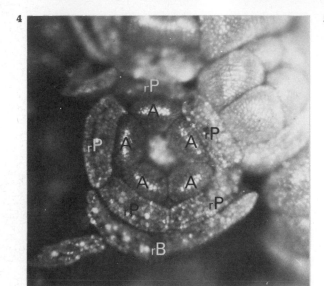

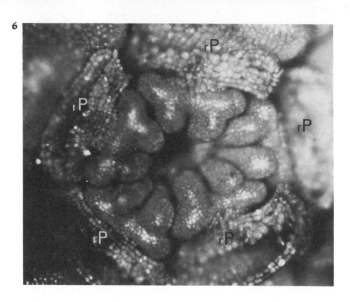

Floral diagram

Floral formula: $\ast$ [P(5) A5 D5]$\bar{\text{G}}$ O2-4

Sequence of primordial inception
P1-5, A1-5, G, D1-5, O1,2,3
Note: At least in some floral buds the tepal primordia (and perhaps the stamen primordia) may be formed in a very rapid succession.

DESCRIPTION OF FLORAL ORGANOGENESIS

The tepal primordia are initiated at about the same time (1, 2). However, in some floral buds one or two of the tepal primordia appear slightly smaller than the others. This small size difference might indicate that at least in these buds the tepal primordia are formed in a very rapid succession. The perianth tube, i.e. growth between the tepal primordia, is initiated immediately before the inception of further primordia (2).

The stamen primordia arise at about the same time at the adaxial base of the perianth tube (3). Since the floral apex has become concave at this stage of development, floral apex and perianth tube are continuous and cannot be sharply delimited from each other. Consequently, the position of the incipient stamen primordia may be termed 'on the floral apex' or 'on the incipient perianth tube,' depending on the arbitrary limit one draws between these two continuous regions. During the further development of the androecium and perianth (4, 5, 6), the zone below the attachment of the stamen primordia elongates. As a result of this, in the mature flower the stamens are inserted on the perianth tube (24, 25). The anthers are dorsifixed and introrse (6, 24).

After the inception of the androecium, the gynoecium arises as a girdling primordium (7). During its upgrowth it closes on top and may assume various outlines (8, 9); but lobes are never formed on the upper margin of the gynoecial primordium. This margin develops gradually into the slightly expanded, simple stigma on top of a thin style (10, 25). Immediately after the inception of the gynoecial primordium, a zone below it grows upward (7, 11, 12). Epigyny results from this zonal growth.

The disc is initiated in the form of five primordia alternate to and central of the stamen primordia (8). Growth extends between the disc primordia around the inner base of the stamen primordia (8). As a result of this, the disc becomes a star-shaped ridge (8, 9, 10). Some of the disc primordia

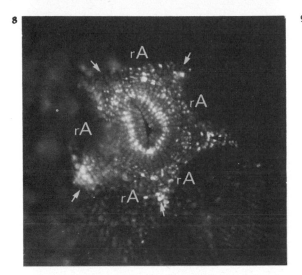

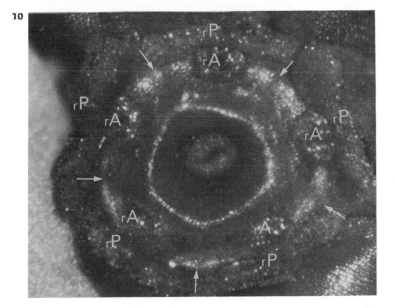

-2 Top views of floral buds with tepal primordia (P) only. The bud in figure 1 is seen from a slight angle. x 146

-6 Top views of floral buds from which tepal primordia were removed (rP) to show inception and development of stamens (A). In figures 5 and 6 the young stamens are not labelled. x 146

-9 Top views of central portion of floral buds showing inception (7) and development of the girdling gynoecial primordium. Stamen primordia were removed (rA). Arrows in figures 8 and 9 point to the incipient disc primordia. x 146

10 Top view of floral bud from which the young stamens (rA) and tepals (rP) were removed. Arrows point to the young disc. Gynoecium is unlabelled. x 85

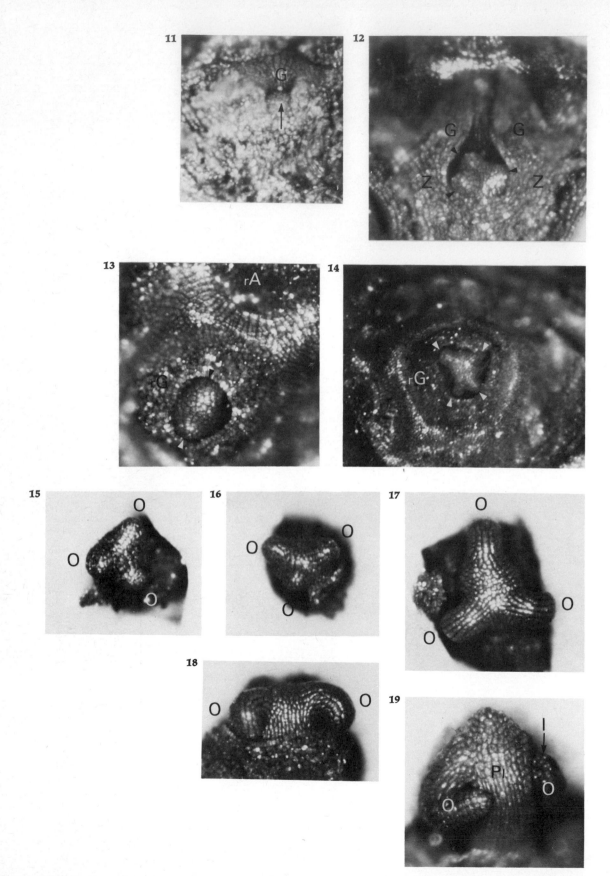

are two-lobed (9). In the mature flower the disc is inserted on the perianth tube (24).

After the inception of the gynoecial primordium, the floral apex resumes growth (11). At this stage, it might as well be called placental apex. At its flanks, it forms successively ovule primordia (12-18). The number of ovule primordia varies from 2 to 4 (or perhaps more). During the inception of the ovule primordia, the floral (placental) apex becomes flat. The ovule primordia first grow toward the base of the ovarial cavity (18). Later on, they grow upward so that the tip of the nucellus points toward the top of the ovary (19-23). During this upward growth of the ovule primordia, the placenta elongates very much and becomes folded and twisted in a very complex manner (19-23). The ovule has one integument which is initiated near the tip of the ovule primordium where it forms a small and inconspicuous rim (19). This rim can only be seen in top views, not in side views (compare, e.g., figure 19 and figure 23).

OTHER AUTHORS

Ram (1957) describes the floral morphology and embryology of *Comandra umbellata* (L.) Nutt. Although he did not study early floral development, he describes the development of the placenta and ovules. His results agree with my observations. His drawings of sections through ovule primordia convincingly show the origin of the distal rudimentary integument. In addition, he illustrates one unusual case in which two ovules were borne on one common stalk (figure 42).

Piehl (1965) published a monograph on the genus *Comandra*. This work includes an extensive bibliography. Ram (1957) quotes papers which deal with the floral development of *Thesium*, another genus of the Santalaceae.

Fagerlind (1948) deals with the morphology of the gynoeceum in the whole order of the Santalales.

BIBLIOGRAPHY

Fagerlind, F. 1948. Gynaeceummorphologie und Phylogenie der Santalales Familien. Sv. Bot. Tidskr. 42: 195-229.

Piehl, M.A. 1965. The natural history and taxonomy of *Comandra* (Santalaceae). Mem. Torr. Bot. Club 22: 1-97.

Ram, M. 1957. Morphological and embryological studies in the family Santalaceae. I. *Comandra umbellata* (L.) Nutt. Phytomorphology 7: 24-35.

FIGURES 11-14. Dissected gynoecia showing different stages in the development of pistil wall and placenta.

11 Side view showing placental primordium (arrow) which, at this stage of development, still may be called floral apex. This gynoecium is at a developmental stage slightly older than that of figure 7. x 146

12 Side view of a gynoecium at a developmental stage corresponding to that of figures 8 and 9. Three ovule primordia were initiated (see arrowheads). Zonal growth (Z) leads to the formation of the inferior ovary. x 146

13 Top view of placenta with two ovule primordia (arrowheads). The young ovary wall was removed (rG). x 146

14 Top view of floral bud from which the young perianth, androecium and pistil wall (rG) were removed to show placenta with four ovule primordia (arrowheads). x 85

FIGURES 15-23. Stages in placental and ovular development.

17 Top views of placentae with three ovule primordia (O). x 146

18 Side view of placenta and ovule primordia in the developmental stage of that of figure 17. x 146

23 Side views of placentae with ovules in different stages of development showing the gradual twisting of the elongating placental stalk (Pl) and the curving of the ovules (O). In figure 19 one ovule primordium is seen from such an angle that the rim-like integument (I) is visible. Figure 21 shows a placenta with four ovules, three of which are visible. Figures 19, 20, 23, x 146. Figures 21, 22, x 85.

24 Nearly mature flower split longitudinally to show placenta (Pl) with ovules, and perianth tube (Pt) with disc (D), two stamens (A) and two tepals (P). x ca. 20

25 Mature flower. x ca. 16

37 / *Comandra umbellata*

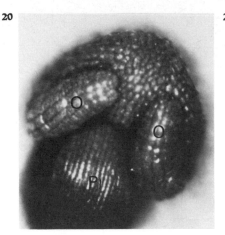

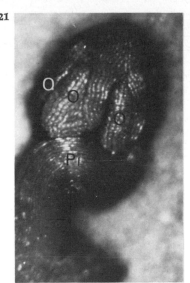

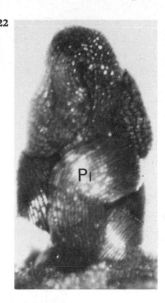

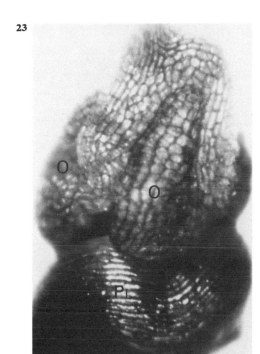

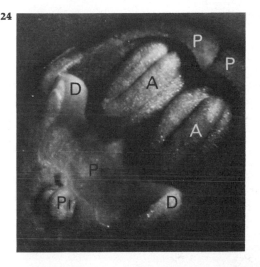

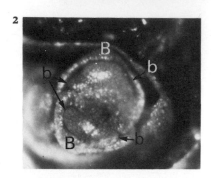

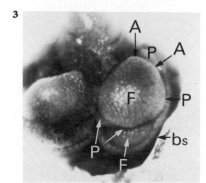

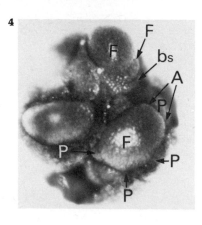

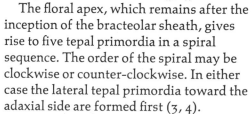

8 / POLYGONALES

POLYGONACEAE

Fagopyrum sagittatum Gilib.
(=**Fagopyrum esculentum** Moench.)
(buckwheat)

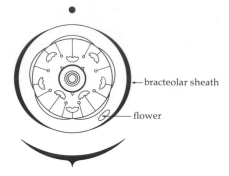

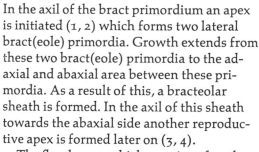

— bracteolar sheath

— flower

Floral diagram

Floral formula: $*[P(5) A_5 + 3$ Glands 8$]$ $\underline{G}(3)$ O1

Note: c = gland

Sequence of primordial inception
P1,2,3,4,5 A_o1-2,3-4,5 A_i1,2,3, G1-3, O1, Glands 1-8
See text for two questionable points in this sequence.

DESCRIPTION OF FLORAL
ORGANOGENESIS

In the axil of the bract primordium an apex is initiated (1, 2) which forms two lateral bract(eole) primordia. Growth extends from these two bract(eole) primordia to the adaxial and abaxial area between these primordia. As a result of this, a bracteolar sheath is formed. In the axil of this sheath towards the abaxial side another reproductive apex is formed later on (3, 4).

The floral apex, which remains after the inception of the bracteolar sheath, gives rise to five tepal primordia in a spiral sequence. The order of the spiral may be clockwise or counter-clockwise. In either case the lateral tepal primordia toward the adaxial side are formed first (3, 4).

Immediately after the inception of the fifth tepal primordium (or even simultaneously with it) a pair of stamen primordia is initiated opposite each of the first two tepal primordia. Very little growth occurs between the two stamen primordia of each pair (3-7). Consequently, the two stamen primordia of one pair seem to be formed on one common primordium. However, it was not possible to observe clearly one single primordium before the appearance of the paired stamen primordia. Instead of referring to two pairs of stamen primordia, one could say that four stamen primordia are alternating with the first two tepal primordia. But one would have to add that the alternation is not quite regular, because the two stamen primordia on either side of a tepal primordium are closer together than the two alternating with the adaxial tepal primordium. The two stamen primordia adjacent to the adaxial tepal primordium are slightly larger than the other stamen primordium of each pair (4-7). The fifth stamen primordium of the outer whorl appears in abaxial position after the two pairs of outer stamen primordia (5). The first stamen primordium of the trimerous inner whorl is initiated adaxially immediately after the fifth one of the outer whorl (or even before the fifth outer one or simultaneously with it). The two other stamen primordia of the inner whorl are the last ones to be formed: the one opposite the third tepal primordium is initiated slightly before the one opposite the fifth tepal primordium (5, 6). During their further development the stamen primordia become first two-lobed (8-10) and then four-lobed as a result of pollen sac formation (11-13). Anther attachment becomes versatile (13). With respect to the orientation of the anther there is a difference between the outer and inner stamens: the anthers of the outer stamens are introrse, whereas the anthers of the inner stamens are extrorse (see Floral Diagram, and figure 13).

After the inception of the inner stamen primordia, the floral apex is flat. It resumes growth and becomes dome-shaped again. At the same time three gynoecial primordia become apparent alternating with the three inner stamen primordia. Immediately afterwards, growth extends between the three gynoecial primordia (14). The gynoecial primordia form the styles and stigmas (15, 20) whereas the cylindrical portion at their base develops into the ovary wall. The closure of the ovary at its tip occurs by growth in the three areas between the developing styles (18). The ovule is formed by continued growth of the floral apex, i.e. the floral apex becomes transformed into the ovule primordium (14). Two integuments are initiated near the base of the ovule primordium, first the inner one (16), then the outer one (17).

While the gynoecium develops, growth occurs at the base of the perianth and androecium including the areas between the perianth members. This leads to the formation of a perigynous flower with an inconspicuous perianth tube. On the developing

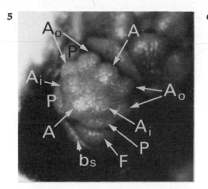

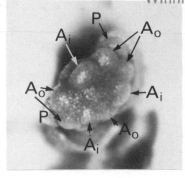

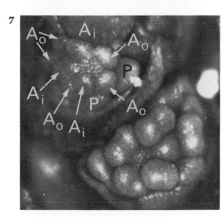

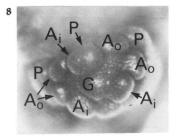

1 Side view of tip of young partial inflorescence showing one apex before and one after inception of bracteoles (b). x 146

2 Top view of two apices with bracteole primordia (b). x 146

-4 Floral buds showing inception of tepals (P) and outer stamen pairs (arrows). A primordial floral apex (F) is visible in the axil of the bracteolar sheath (bs). Both photographs show the same buds in different focus and angle. x 146

-6 Different angles of floral buds with the primordia of tepals and outer and inner stamens (A_o and A_i). x 146

7 Two floral buds, one of them labelled. x 146

10 Top views of floral buds showing inception and early development of gynoecium (G). x 146

11 Adaxial view of a floral bud. x 146

12 Nearly abaxial view of a floral bud in a similar developmental stage as that of figure 11. x 146

13 Side view of one inner stamen (A_i) and two outer stamens (A_o). x 85

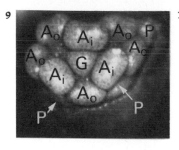

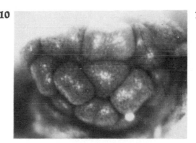

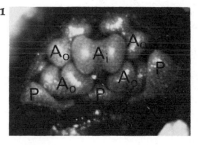

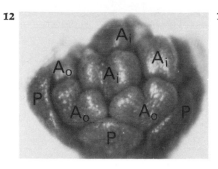

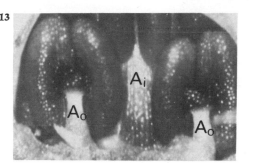

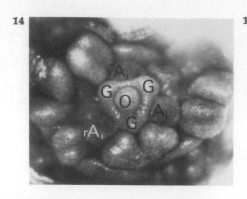

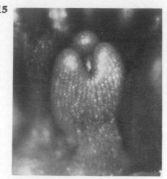

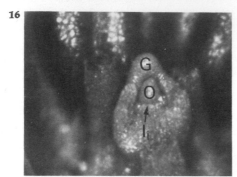

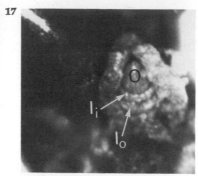

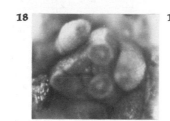

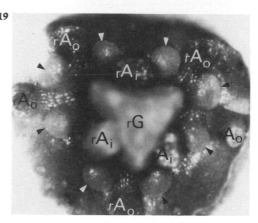

perigynous cup eight primordia are initiated which will give rise to glands (19).

Two types of flowers are found in this species. One type has styles shorter than stamens (20, 21); in the other type the styles are longer than the stamens (22). These differences develop late during floral ontogeny.

OTHER AUTHORS

No detailed account of the floral organogenesis of *Fagopyrum esculentum* has been published to my knowledge. Bauer (1922) describes the floral development of the genus *Polygonum*, which has the same floral diagram as *Fagopyrum*. He mentions that after the spiral inception of the five tepals the five outer stamens are initiated 'almost simultaneously.' The three inner stamens are said to appear after the outer ones in the same sequence as reported here. His findings on pistil development also agree with mine.

According to Schumann (1890) who studied *Polygonum bistorta* the adaxial inner stamen primordium and the adaxial tepal primordium are initiated after the first two tepal primordia. At about the same time two primordia are formed superposed to the first two tepal primordia. On each of these primordia two stamen primordia are initiated. After this formation of primordia on the adaxial side of the floral bud, a tepal primordium is formed toward the abaxial side, adjacent to the first tepal primordium and, simultaneously, a primordium opposite this tepal primordium. The latter primordium gives rise to the fifth outer stamen primordium and the second inner one. Finally, the fifth tepal primordium is formed towards the abaxial side of the bud and the third inner stamen primordium superposed to it. To express this sequence in one technical term, it could be called 'absteigend' (descendent) with the exception of the adaxial tepal. 'Absteigend' (descendent) means that the primordial inception starts at the adaxial side of the floral apex and proceeds gradually towards its abaxial side. My observations do not confirm Schumann's, but perhaps the adaxial inner stamen primordium appears before the abaxial outer stamen primordium as postulated by Schumann.

Payer (1857) describes the floral development of two species of *Polygonum*. His findings are very similar to mine with one exception: Payer claims that each of the two stamen pairs originates from one common primordium. However, Payer did not draw a developmental stage which shows the common androecial primordia before they produce the pairs of stamen primordia. Payer drew a slightly later stage which corresponds to my figure 5.

A number of authors such as Emberger (1939), Geitler (1929), Jaretzky (1928), Laubengayer (1937) studied the mature flowers of the Pologonaceae. All of these publications are quoted by Vautier (1949). Schoute (1932) made a detailed study of variation in the polygonaceous flower. Among many other changes in floral structure, he found single stamens instead of the pairs of stamens opposite the first two tepals. Mahoni (1935) and others describe the embryology of *Fagopyrum esculentum*. For literature on heterostyly in *Fagopyrum* see Schoch-Bodmer (1930).

BIBLIOGRAPHY

Bauer, R. 1922. Entwicklungsgeschichliche Untersuchungen an Polygonaceenblüten. Flora *115*: 273-292.

Mahoni, K.L. 1935. Morphological and cytological studies on *Fagopyrum esculentum*. Amer. J. Bot. 22: 460-475.

Payer, J.B. 1857. *Traité d'organogénie comparée de la fleur. Texte et Atlas*. Paris: Librairie de Victor Masson.

Schoch-Bodmer, H. 1930. Zur Heterostylie von *Fagopyrum esculentum*. Bull. Soc. bot. Suisse *39*: 4.

Schoute, J.C. 1932. On pleiomery and meiomery in the flower. Rec. trav. bot. néerl. *29*: 164-226.

Shumann, K. 1890. *Neue Untersuchungen über den Blütenanschluss*. Leipzig.

Vautier, S. 1949. La vascularization florale chez les Polygonacées. Candollea *12*: 219-341.

20

21

22

FIGURES 14-17. Stages in ovule development. x 146

14 Top view of young gynoecium with three gynoecial primordia (G) and the ovule primordium (O). Inner stamen primordia were removed (rA_i).

15 Side view of young gynoecium. The tip of the ovule primordium is just visible.

17 Dissected young pistils showing ovule primordia with incipient inner (I_i) and outer integument (I_o).

18 Top view of young pistil showing the three style primordia. x 85

19 Floral bud from which gynoecium, tepals, and stamens were removed to show the eight gland primordia (see arrowheads). x 146

FIGURES 20-22. Mature flowers. x ca. 7

21 Flowers with short styles and long stamens.

22 Flower with long styles and short stamens.

9/CENTROSPERMAE

CARYOPHYLLACEAE

Silene cucubalus Wibel
(bladder-campion)

Floral diagram

Floral formula: ✳ K(5) C5 A(5+5) G̲(3) O∞

Sequence of primordial inception
K1,2,3,4,5, C1,2,3,4,5, A$_k$1,2,3,4,5, A$_c$1,2,3,4,5, G1-3,
O many in basipetal sequence
Note: The primordia of the petals and the two sets of
stamens are formed in a very rapid succession which
approaches simultaneous origin.

DESCRIPTION OF FLORAL ORGANOGENESIS

The sepal primordia are formed in a rapid
spiral sequence. The first three sepal primordia are larger than the last two (1), i.e.
there is a well pronounced plastochron between the inception of the third and fourth
sepal primordia. The plastochron between
the inception of all the other sepal primordia is minimal, if it exists at all. Immediately
after the inception of the sepal primordia,
the calyx tube is initiated due to growth in
the regions between the sepal primordia.
This tube grows more than the original
sepal primordia, which are carried up on it.
Consequently, in the mature flower the
calyx tube is much longer than the calyx
lobes.

Soon after the inception of the calyx
tube, the petal primordia are initiated at
about the same time or in a very rapid succession which is difficult to ascertain (2, 3).
They grow very slowly compared to all
other floral organs (6, 7, 18, 19, 20), and
finally become two-lobed (20, 22, 24). Two
little teeth (paracorolla) develop late adaxially below the two lobes of each petal.

The antesepalous stamen primordia are
initiated immediately after the petal primordia in a very rapid succession which
approaches simultaneity (4). Thereafter,
also in a very rapid succession, the antepetalous stamen primordia arise near the
adaxial base of the petal primordia (4, 6).
Since the adaxial base of the petal primordia is continuous with the floral apex, the
two regions cannot be sharply delimited
from each other. Consequently, the decision
whether the antepetalous stamen primordia
arise on the adaxial side of the petal primordia or on the floral apex becomes a matter of drawing an arbitrary limit. The inner
margins of the antepetalous stamen primordia are at about the same level as those of
the antesepalous stamen primordia (7).
During inception of the antepetalous
stamen primordia, their inner margins are
even slightly outside (i.e. at a lower level
than) those of the antesepalous stamen primordia (4). The outer (abaxial) margins of
the antepetalous stamen primordia are always inside (i.e. at a higher level than)
those of the antesepalous stamen primordia
(4, 6, 18). Thus, with respect to the outer
limit of stamen insertion, the androecium
is diplostemonous, whereas with regard to
the inner limit it is first slightly obdiplostemonous, and then intermediate between
the diplostemonous and obdiplostemonous
condition, since the inner margins of all
stamen primordia and stamens are at about
the same level (7, 21). During the development of the androecium, the antepetalous
stamen primordia are smaller than the antesepalous ones (4, 7, 9, 18, 19). Some growth
occurs in later stages of development in a
zone below the base of all stamen primor-

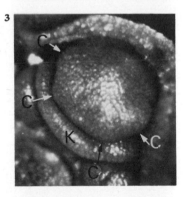

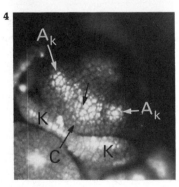

dia. This leads to a common base (21) which surrounds the gynophor in the mature flower. Considerable growth occurs below each petal primordium and its superposed stamen primordium. This growth is already noticeable in very early stages of development (6), but becomes more prominent in later stages and leads finally to the insertion of the antepetalous stamens high up on the petals (23, 24). All the stamen primordia develop dorsifixed, introrse anthers. Two types of flowers occur on different plants: in one type the stamens are much shorter than the calyx tube (24), whereas in the other type the stamens are longer than the calyx tube (22, 23).

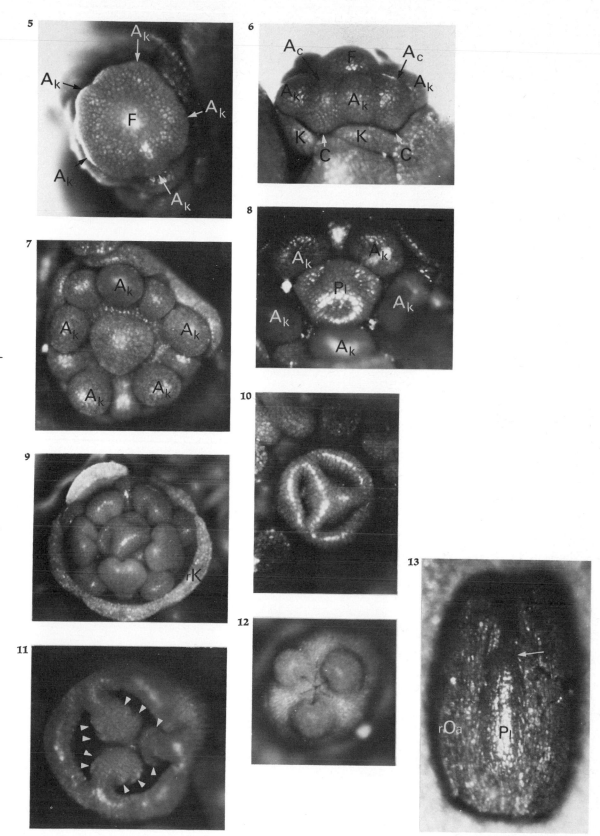

FIGURES 1-8. Top and side views of floral buds, showing inception and early development of sepals, petals, stamens, and gynoecium. x 146

1 Floral apex (F) with three sepal primordia (K).
2-3 Inception of calyx tube and petals (C).
4 Inception of one antepetalous stamen (arrow) between two antesepalous stamen primordia (K).
5 Arrows point to the five antesepalous stamen primordia.
6 The two small antepetalous stamen primordia between the antesepalous ones (A_k) are labelled A_c.
7 Gynoecial inception. Only the antesepalous stamen primordia are labelled (A_k).
8 Beginning of locule formation, i.e. inception of pistil wall, three septa, and central column (Pl).
9 Floral bud from which the upper portion of the young calyx was removed (rK) to show development of stamens and gynoecium. x 85
11 Top view of young gynoecium. In figure 11 ovule inception (see arrowheads) can be seen on the septa which extend above the central column. x 146
12 Top view of young pistil after closure of ovary. Note the three style primordia which are visible already in figures 7 and 8. x 85
13 Side view of young pistil in a developmental stage between that of figures 10 and 11. Part of the wall was removed (rOa) to show the central column (P1) whose upper end is marked by an arrow. x 146

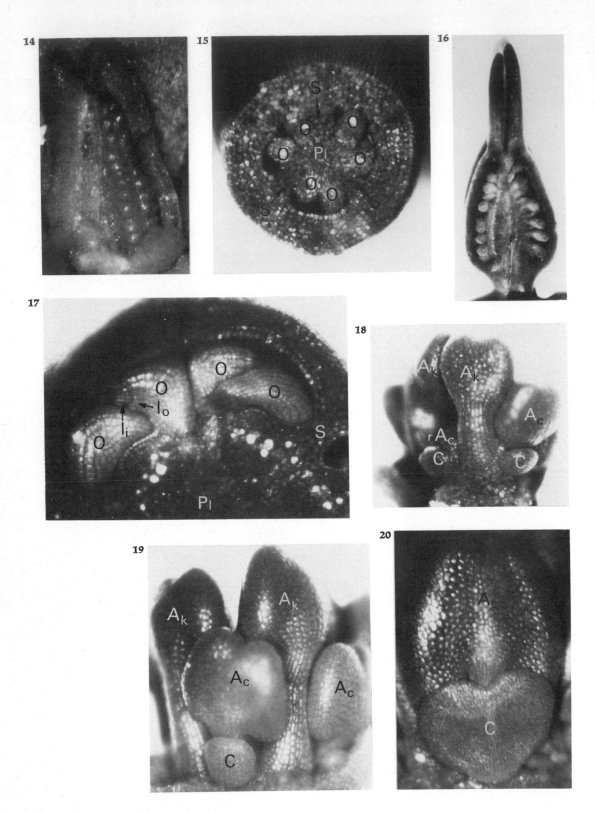

During and after the inception of the petal and stamen primordia the floral apex has become discretely dome-shaped (2, 6). Then it assumes a triangular shape (7), and growth becomes very much restricted in three areas, the future locules (8-10). Eventually growth becomes retarded in the center of the gynoecium. The septa, i.e. the growing regions which separate the locules, grow up farther in continuity with the ovary wall (11). The three portions of the ovary wall opposite the locules grow faster and close the ovary on top with their margins (12); their terminal extensions will form the three styles and stigmas (16). Yet before the closure of the ovary, ovule primordia are initiated basipetally in three double rows on the central column (placenta) and on the septa above this column (14, 15). Each ovule develops two integuments (17).

The gynophor (23) is initiated already during the inception of the gynoecium (7, 8). Like an internode, it elongates gradually during the further development of the flower.

OTHER AUTHORS

Rohweder (1967) described the floral development of *Silene dioica* and other members of the Silenoideae. The main features of development of *Silene dioica* are the same as in *Silene cucubalus*. Moeliono (1970) investigated gynoecial development in *Silene alba* and other members of the Caryophyllaceae. He found that the septa fuse postgenitally with the central column. No evidence for such fusion was detected in this study. As Rohweder (1967), Hartl (1956, quoted by Moeliono), Lister (1883, quoted by Moeliono) and others noted there is continuous upgrowth of the septa with the central column. Kraft (1917, quoted by Rohweder, 1967) considers the androecium of *Silene venosa* as slightly obdiplostemonous, although he points out that all stamens are inserted at about the same level. Thompson (1942) who studied the vascularisation of mature flowers in five other species notes that the traces of all ten stamens arise in one whorl. Schumann (1890) studied mainly the early stages of floral organogenesis in *Silene armeria*. He concludes that the first three sepal primordia appear at the same time and are followed by the simultaneous inception of the fourth and fifth sepal primordia. This agrees roughly with my observations.

Eckert (1965) described the floral development of *Cerastium chlorifolium* and *Spergula arvensis*. Roth (1963) reported the development of the gynoecium in the genus *Cerastium* which differs from that of *Silene* in the number of locules and relative growth rates.

BIBLIOGRAPHY

Eckert, Gertrud. 1965. Entwicklungs-geschichtliche und blütenanatomische Untersuchungen zum Problem der Obdiplostemonie. Bot. Jb. *85*: 523-604.

Hanzel, R.J. et al. 1955. Floral initiation and development in the carnation, var. Northland. Proc. Amer. Soc. Hort. Sci. *6*: 455-462.

Moeliono, B.M. 1970. *Cauline or carpellary placentation among dicotyledons*. I. Text. II. Plates. Assen (Netherlands): Van Gorcum and Co.

Rohrbach, P. 1868. Monographie der Gattung Silene (cited in: Thompson, Betty F. 1942).

Rohweder, O. 1967. Centrospermen-Studien. 3. Blütenentwicklung und Blütenbau bei Silenoideen (Caryophyllaceae). Bot. Jb. *86*: 130-185.

Roth, I. 1963. Histogenese und morphologische Deutung der Zentralplazenta von *Cerastium*. Bot. Jb. *82*: 100-118.

Schumann, K. 1890. *Neue Untersuchungen über den Blütenanschluss*. Leipzig: W. Engelmann.

Thompson, Betty F. 1942. The floral morphology of the Caryophyllaceae. Amer. J. Bot. *29*: 333-349.

14 Similar dissection of a pistil as that in figure 13, but in a later stage after ovule inception. x 85

15 Cross-section through young ovary showing the three septa (S) and ovule primordia (O) on the central column (P1). x 146

16 Nearly mature pistil from which part of the ovary wall was removed. x ca. 20

17 Portion of a cross-section through an ovary showing one septum (Se) and ovules (O) on the central column (Pl) with two integuments (I_1, I_0). x 146

18 Side view of a floral bud after removal of calyx. One antepetalous stamen primordium was removed (rA_c). The primordia of two petals (C), two antesepalous (A_k) and one antepetalous stamen (A_c) are visible. x 146

19 Slightly older stage of development than that of figure 18. x 146

20 Two-lobed petal primordium (C) and antepetalous stamen (A_c). x 85

21 Inner side of androecial ring. Pistil was removed (rG). x 85

22 Mature flower. x ca. 2

24 Side views of mature flowers from which part of the perianth was removed to show the long stamens in figure 23, and the short stamens in figure 24. x ca. 2

22

23

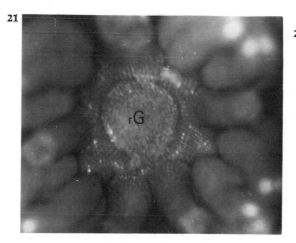

21

24

10/CENTROSPERMAE

CHENOPODIACEAE

Chenopodium album L.
(white goosefoot)

Floral diagram

Floral formula: $* P_5 A_5 \underline{G}(2) O_1$

Sequence of primordial inception
$K_{1,2,3,4,5}$, $A_{1,2,3,4,5}$, $G_{girdling}$, G_{1-2}, O_1
Note: The ovule primordium is the continuation of the floral apex. It is not initiated as a separate entity.

DESCRIPTION OF FLORAL ORGANOGENESIS

The primordia of the five perianth members are initiated in a spiral sequence (1-5). In some floral buds the plastochron between successive primordia appears to approach zero (1), whereas in others it is conspicuous (3)

The five stamen primordia are initiated opposite the five primordial perianth members and in the same sequence as the latter (3, 4), i.e. the first stamen primordium appears in the axil of the first primordial perianth member, the second stamen primordium in the axil of the second primordial perianth member, and so on. All stamen primordia grow very slowly compared to all other floral organs including the gynoecium (9). They develop quite suddenly in the late stages of floral development. Each primordium forms an introrse stamen. Little growth occurs in the region between the stamens which leads to the formation of a rudimentary androecial tube (13). This tube is hardly noticeable in the mature flower and, therefore, has not been indicated in the floral diagram.

After stamen inception, the gynoecium is initiated in the form of a girdling primordium (5). As this gynoecial primordium grows up, its rim becomes two-lobed, i.e. two gynoecial primordia form on the girdling gynoecial primordium (6-8). These two gynoecial primordia develop into stylar branches with stigmas, whereas the girdling gynoecial primordium gives rise to the ovary and the common style (9-11, 14).

The floral apex becomes gradually transformed into the single ovule (5, 6). As it develops, it forms two integuments and becomes anatropous (10, 11, 12).

FIGURES 1-2. Floral apices with the primordia of perianth members (P). x 146

1 Top view.
2 Side view of another bud.
3 Side view of floral bud with primordia of perianth members (P) and stamens (A). x 146
4 Top view of a floral bud from which the primordia of one perianth member and its superposed stamen were removed (rP and rA resp.). Among the four stamen primordia (A) present, two of them are clearly visible, one is partially hidden and one is completely hidden by its subtending primordial perianth member. x 146
5 Top view of floral bud showing girdling gynoecial primordium. x 247
8 Floral buds showing formation of two gynoecial primordia on the girdling primordium. x 146
9 Side view of floral bud from which the young perianth was removed to show the retarded stamen primordia (A) and pistil (G). x 146
1 Young pistil from which a portion of the ovary wall was removed to disclose the young ovule with integument primordia. Figure 10, x 247. Figure 11, x 146
2 Nearly mature ovule. x 146
3 Old floral bud from which pistil and perianth was removed to show the inconspicuous upgrowth between the bases of the stamens, thus joining all stamens on a common ring. x 85
4 Mature flower. x ca. 30

OTHER AUTHORS

Cohn (1914) gave a brief description of floral development in the genus *Chenopodium*. My observations agree with his account. Lundblad (1922) made a detailed study of variation in the number and position of floral organs in *Chenopodium album* and *Ch. rubrum*. Like Cohn (1914) and Lundblad (1922), Eckardt (1967b) pointed out that considerable variation in floral structure may be found even on the same inflorescence. He mentioned that a more detailed study of this variation will be undertaken by Mr M. Hakki. I did not study the variation within one inflorescence. My observations are based only on the terminal flower buds of the main inflorescence branches. All of these buds showed the configuration presented in the floral diagram.

Eckardt (1967a, b) described the floral devolpment in the closely related genus *Dysphania* which shows different patterns of floral construction that may also occur in the genus *Chenopodium*.

BIBLIOGRAPHY

Cohn, F.M. 1914. Beiträge zur Kenntnis der Chenopodiaceen. Flora *106*: 51-89.
Eckardt, T. 1967a. Blütenbau und Blütenentwicklung von *Dysphania myriocephala* Benth. Bot. Jb. *86*: 20-37.
– 1967b. Vergleich von *Dysphania* mit *Chenopodium* u. mit Illecebraceae. Bauhinia *3*: 327-344.
Lundblad, H. 1922. *Über die baumechanischen Vorgänge bei Entstehung von Anomomerie bei homochlamydeischen Blüten*. Thesis. Lund.

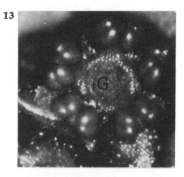
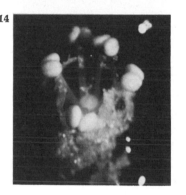

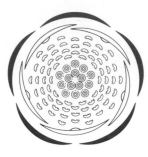

11 / RANUNCULALES

RANUNCULACEAE

Ranunculus acris L. (buttercup)

Floral diagram

Floral formula: $*$ K5 C5 A∞ G∞ O∞
Note: A nectariferous scale is at the adaxial base of
the petal. The gynoecium is apocarpous with one
ovule per pistil.

Sequence of primordial inception
K1,2,3,4,5, C1-5, A∞ in partial succession, G∞ in
partial succession, each forming one ovule.
Note: 'Partial succession' means that the primordia
overlap in timing of their origin, i.e. several of the
primordia at one level of the floral dome appear more
or less at the same time.

DESCRIPTION OF FLORAL ORGANOGENESIS

The sepal primordia are initiated in a spiral
sequence at the base of a high dome-shaped
apex (1). Shortly after the inception of the
fifth sepal primordium, the five petal primordia appear at about the same time (2).
The possibility, however, that they are
formed in a very rapid spiral succession cannot be excluded. At first the petal primordia
have the same shape as the stamen primordia (3, 4, 5). Later on, marginal growth
occurs and the primordia become more leaf-
like (18). At that time an elongate primordium is initiated at the adaxial base of
the petal primordium (19). This primordium
develops into the nectariferous scale (20). In

comparison with the development of the
sepals, stamens, and pistils, growth of the
petal primordia is very retarded. Even at a
stage in which the four pollen sacs of stamens have been initiated, the petal primordia are smaller than the stamen primordia (18). The rapid elongation of the petal
primordia occurs shortly before the flower
opens (21).

The first stamen primordia are initiated
more or less simultaneously immediately
after the petal primordia, or, in some buds,
perhaps even overlapping with petal inception (2, 3). The position of these first stamen
primordia is somewhat opposite the sepal
primordia. However, more than one stamen
primordium may be initiated opposite a
sepal primordium. The stamen primordia
formed next alternate with the first ones, and
so on. As a result of this continued alternation, vertical and horizontal rows of stamen
primordia become apparent (4-9). However,
irregularities are quite common. Sometimes
a vertical row comprises only two primordia
counting from the base of the sepal or petal
primordia; sometimes alternation is not
well pronounced. Due to these and other
irregularities it may be difficult to establish
the total number of rows for a floral bud.
Keeping this in mind, it may be said that the
number of vertical rows varied from 14 or
15 to 25 or 26, whereas the number of horizontal rows varied from 3 to 5 in the material examined. Probably more variation
could be found, if more buds were studied.
One bud was found with 6 sepals and 6
petal primordia; of course, this bud had a
larger number of vertical stamen rows (7).
In general, the distribution of vertical rows
is not quite symmetrical with respect to the
sepal primordia. Since the first sepal primordium is initiated much earlier than the
fifth one, it is much larger than the fifth one
at the time of stamen inception. Consequently, there is more space available opposite the first sepal primordium than opposite
the fifth one. This may explain why usually

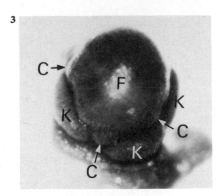

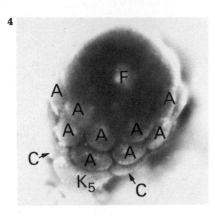

only one (two, or three) row(s) of stamen primordia are formed opposite the fifth sepal primordium (3, 4), whereas (three), four, or five (or more) staminal rows occur opposite the first sepal primordium (6). The number of stamen rows opposite the other sepal primordia varies between these extremes. During their further development, the stamen primordia elongate in lateral direction (8, 9) and finally become differentiated into filaments and dorsifixed, extrorse anthers (10, 18, 21, 22).

The sequence and position of pistil inception parallels that of the stamens, i.e. pistil primordia appear in more or less distinct vertical and horizontal rows. The number of vertical pistil rows is not necessarily the same as that of the stamens. But often the pistil primordia continue the vertical stamen rows (9, 10). The shape of the incipient pistil primordia is the same as that of the

---

FIGURES 1-4. Floral buds showing inception of sepals, petals, and first stamens. x 146

1 Side view of dome-shaped floral apex (F) showing two sepal primordia (K) in different stages of development.
2 Primordia of two sepals (K), three petals (C), and the first stamens (A).
3 Similar stage as that in figure 2.
4 Fifth sepal primordium (K₅) with the adjacent petal and stamen primordia. One vertical row of stamen primordia is opposite the fifth sepal primordium.
5 Floral bud showing three rows of stamen primordia opposite the fifth sepal primordium (K₅). x 85
6 Top view of unusual bud with six sepal and six petal primordia. The number of vertical stamen rows is indicated opposite an outer and inner sepal primordium (K). x 146
-8 Top (8) and side (7) view of one floral bud after stamen inception. Sepal primordia were removed. x 85
9 Side view of floral bud showing inception of the first gynoecial primordia (G) above stamen primordia (A). x 146
o Top view of floral bud showing stamen primordia and gynoecial primordia in different stages of development. x 85

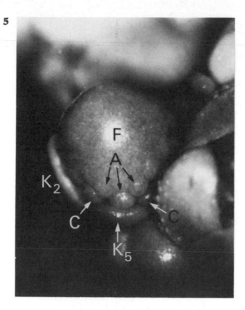

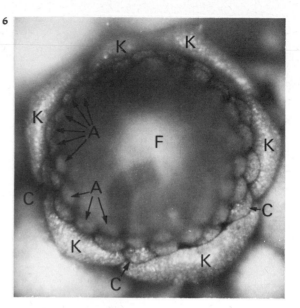

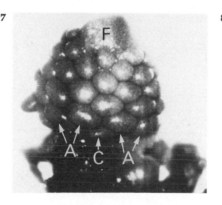

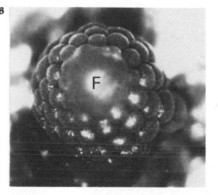

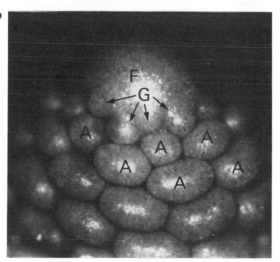

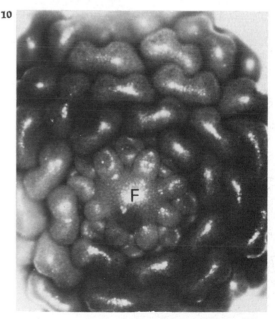

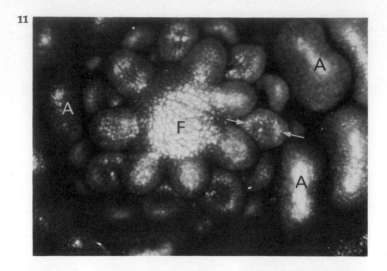

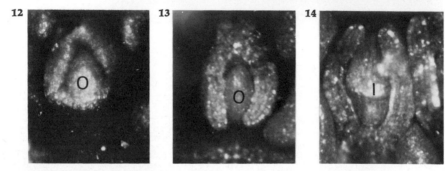

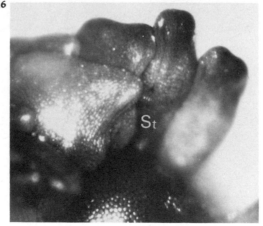

young stamen primordia. However, soon after the inception of the pistil primordia, drastic changes occur. The central apical portion of each pistil primordium stops growing or is at least very retarded in its growth, whereas the periphery of the primordium grows considerably, thus forming a rim around a central 'depression' (10, 11). Then two growth centers become distinguishable on the incipient rim (10, 11): an abaxial horseshoe-shaped one which will form a hood-like structure (12, 15, 16), and an adaxial roundish one which will give rise to the ovule (12, 13). The hood-like structure grows at its margins until they become postgenitally fused. At the same time it expands into a lower portion, the ovary, which encloses the single ovule, and an upper narrow portion, the style (17). The stigma differentiates along the fused margins at the tip of the style. On the ovule primordium (12, 13) one integument is formed (14). The mature ovule becomes hemianatropous. The common base from which ovary wall and ovule developed (often interpreted as the stalk of a 'peltate carpel') does not elongate much. However, one pistil was found in which this portion elongated considerably thus forming a stalked pistil (16). At the adaxial base of the ovule, where the margins of the ovary wall fuse, an inconspicuous bulge grows up in continuity with the bases of the fusing margins. This bulge (in interpretive language often called 'querzone') is initiated long after the ovule at the base of the latter (12, 15, 16).

OTHER AUTHORS

I found no literature on the floral organogenesis of *Ranunculus acris*. However, other species of the genus *Ranunculus* were described in great detail. Tepfer (1953) made an excellent study of the floral development of *Ranunculus repens*. His observations agree with mine on *Ranunculus acris* with one main exception; Tepfer (p. 554) wrote:

'The stamens are inserted in spirals, clearly not in whorls, and they do not form distinct vertical columns as in *Aquilegia*.' This is not quite true for *Ranunculus acris*. No distinct spiral arrangement of the stamens could be detected. In fact, the arrangement of the stamens comes rather close to a whorled condition with some irregularities. Thus, the position of the stamens in *Ranunculus acris* seems to be somewhat intermediate between that in *Ranunculus repens* and *Aquilegia formosa* as described by Tepfer. Hiepko's (1965) report on perianth development in the genus *Ranunculus* agrees with my observations; so do Rohweder's (1967) and Eckardt's (1957, quoted in Rohweder 1967) results on gynoecial development. The major difference between Payer's (1857) description of the floral organogenesis in *Ranunculus trilobus* and my observations concerns the number of stamens, which is smaller in *Ranunculus trilobus*.

17

18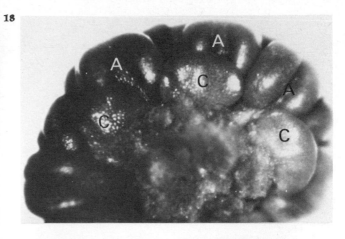

20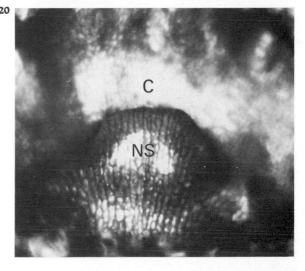

19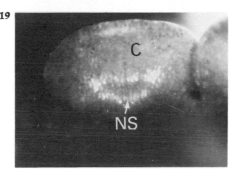

21

22

---

11 Same stage of development as in figure 10. On one of the gynoecial primordia the two growth centers are indicated by arrows. x 146

12 Adaxial view of one young pistil showing ovule primordium (O) and hood-like structure which will form pistil wall. x 146

14 Young pistils from which the upper portion was removed to show the ovule primordium before (13) and after (14) inception of the integument (I). x 146

15 Adaxial view of two young pistils. From the one to the right the wall was removed to disclose the ovule primordium (O). x 146

16 Three young pistils; one of them is unusual in that it has a stalk (St). x 85

17 Top view of young pistils after formation of styles. x 146

18 Half of a floral bud seen from below to show retardation of petal primordia (C) in comparison to the stamen primordia (A). x 85

19 Upper side of a petal primordium showing inception of the nectariferous scale (NS). x 146

20 Nectariferous scale in a later stage of development. x 146

21 Top view of old bud after removal of sepals showing petals and stamens. x ca. 18

22 Mature flower. x ca. 2.5

In *Ranunculus pallasii*, Leinfellner (1958) reported a morphological series of nectariferous glands ranging from a structure similar to the one reported here to those which are continuous with the margin of the petal and thus produce a peltate-ascidiate petal. Leinfellner, however, did not describe the development of these glands.

BIBLIOGRAPHY

Fromantin, Jane. 1956. Quelques considérations d'ordre évolutif à la suite des effects du 2,4-D sur la structure des carpelles de *Ranunculus arvensis* L. Rev. Gén. Bot. *63*: 293-313.

Hiepko, P. 1965. Vergleichend-morphologische und entwicklungsgeschichtliche Untersuchungen über das Perianth bei den Polycarpicae. Bot. Jahrb. *84*: 359-426.

Leinfellner, W. 1958. Beiträge zur Kronblattmorphologie. VIII. Der peltate Bau der Nektarblätter von *Ranunculus*, dargelegt an Hand jener von *Ranunculus pallasii* Schlecht. Österr. bot. Zschr. *105*: 184-192.

Payer, J.B. 1857. *Traité d'organogénie comparée de la fleur. Texte et Atlas.* Paris: Librairie de Victor Masson.

Rohweder, O. 1967. Karpellbau und Synkarpie bei Ranunculaceen. Ber. Schweiz. Bot. Ges. *77*: 376-432.

Tepfer, S.S. 1953. Floral anatomy and ontogeny in *Aquilegia formosa* var. *truncata* and *Ranunculus repens*. Univ. Calif. Publs. Bot. *25*: 513-648.

12 / PIPERALES

PIPERACEAE

Peperomia caperata Yuncker

Floral diagram

Floral formula: $\cdot | \cdot$ A2 $\underline{G}$(2) O1

Sequence of primordial inception
A1-2, $G_{girdling}$, G1-2, O1

DESCRIPTION OF FLORAL ORGANOGENESIS

During a process of lateral expansion the floral apex forms two stamen primordia in the transversal plane (1). In most buds one of the two primordia is larger than the other even at a rather early stage (1, 2). Each of them becomes two-lobed, and eventually develops into a stamen with a bilocular anther which is obliquely oriented on the filament (3-6).

After the stamen primordia have become quite prominent (2), growth occurs in a position between the stamen primordia, but slightly adaxial of the transversal plane. The result of this growth is a primordium which may be called the florax apex or gynoecial primordium. Soon after its inception, maximal growth shifts from its center to its periphery. The result of this is the inception of a girdling primordium (3). As this primordium grows upward, thus forming the pistil wall, it develops two very inconspicuous lobes in the median plane (4, 5, 6). These two lobes become appressed towards each other and close the pistil in this manner (7). Usually the abaxial lobe is slightly more developed than the adaxial one. Each of them differentiates into a stigma (9, 10, 12). Occasionally buds were found with three stigmas, due to the formation of three lobes on the girdling gynoecial primordium (11).

The floral apex resumes growth and becomes transformed into an ovule primordium. This primordium forms one integument as it develops into an orthotropous ovule. At least in some cases the position of the ovule may slightly change towards the adaxial side of the ovarial base (7, 8).

OTHER AUTHORS

I found no recent literature on the floral development of *Peperomia* spp. Long ago, Schmitz (1872) studied the floral development of almost 30 species of *Peperomia*. Most of his results are in agreement with my observations; however, he found only one stigma. Variation occurred in the lobing of the gynoecial primordium: in some species the adaxial lobe is more developed, in others the abaxial lobe grows more than the adaxial one, and in some species lobing does not occur (p. 29). Like Abele (1923) and in agreement with my observations, Schmitz reported for all species that the ovule arises in an exactly central position,

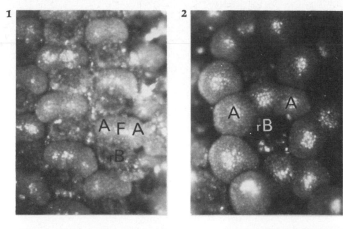

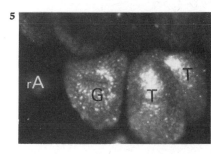

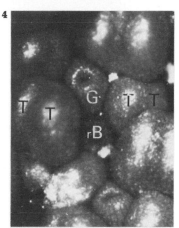

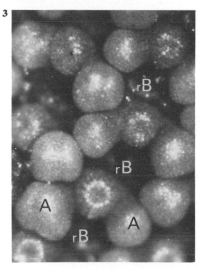

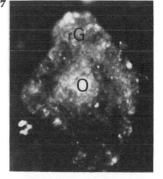

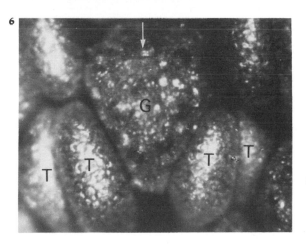

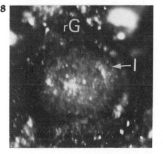

FIGURES 1-4. Portions of young inflorescences; bracts removed (rB). x 146

2 Floral apices with lateral stamen primordia (A).

3 Inception of circular gynoecial primordium.

4 Formation of thecae (T) on the stamen primordia; formation of two gynoecial primordia on the circular gynoecial primordium.

5 Floral bud from which one stamen primordium was removed (rA) showing closure of gynoeceum. x 146

6 More advanced floral bud. Arrow points to the appressed gynoecial primordia. x 146

8 Gynoecia from which the young ovary wall was removed to show inception and development of the ovule. The arrow in figure 8 points to the incipient integument (I). x 146

9

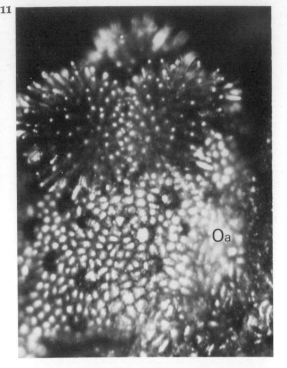

11

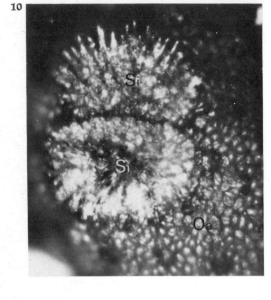

10

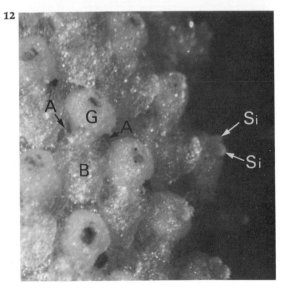

12

i.e. as the continuation of the floral apex. Baillon (quoted by Schmitz 1872), who studied a number of cultivated *Peperomia* and *Piper* species, came to a different conclusion than I did with regard to ovule position. He stated that the ovule is initiated adaxially near the base of the young ovary. The ovary wall, he wrote, arises as a circular primordium, or as a crescent-shaped primordium on the abaxial side. Murty (1952), who studied only mature flowers, also claimed a sub-basal position of the ovule. Eckardt (1937) described and discussed the morphology and anatomy of mature flowers of *Piper* and *Peperomia*. He quoted Abele (1923), Hagerup (1934), and Johnson (1902) who reported also on floral development of *Peperomia* and/or *Piper*. Johnson (1914) described the floral development and embryology of *Peperomia hispidula*.

BIBLIOGRAPHY

Eckardt, Th. 1937. Untersuchungen über Morphologie, Entwicklungsgeschichte und systematische Bedeutung des pseudomonomeren Gynoeceums. Nova Acta Leopold. N.F. 5: 1-112.

Johnson, D.S. 1914. Studies of the development of the Piperaceae. II. The structure and seed development of *Peperomia hispidula*. Amer. J. Bot. 1: 323-339 and 357-397.

Murty, Y.S. 1952. Placentation in *Peperomia*. Phytomorphology 2: 132-134.

Schmitz, F. 1872. Die Blüthen-Entwicklung der Piperaceen. In: Hanstein's botan. Abhandlungen. Band 2, Heft 8: 1-74.

Yuncker, T.G. 1957. New species in *Peperomia*. Kew Bull. 12: 421-422.

9-10 Gynoecia showing differentiation of the ovary wall and the two stigmas. Figure 9, x 85. Figure 10, x 146

11 Gynoeceum with three stigmas. x 146

12 Portion of mature inflorescence showing several flowers. The bracts (B) obscure part of the stamens (A). x ca. 20

13 / GUTTIFERALES

DILLENIACEAE

Hibbertia scandens (Willd.) Dryand (= H. volubilis Andre)

Floral diagram

Floral formula: $\ast$ K5 C5 A∞ G5 O5 × 4-6

Sequence of primordial inception
K1,2,3,4,5, C1,2,3,4,5, A∞ in centrifugal sequence overlapping in time of origin with G1,2,3,4,5, O5 times 1-4(-6)
Note: Especially the first three petals approach simultaneous inception. Some of the ovules may appear sequentially.

DESCRIPTION OF FLORAL ORGANOGENESIS

The sepal primordia are initiated in a very distinct spiral sequence (1-3). The plastochron between the inception of successive sepal primordia is long; as a result of this, the first sepal primordium is quite large when the fifth one is initiated. It is difficult

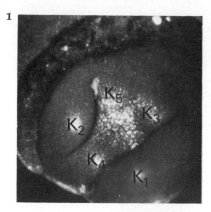

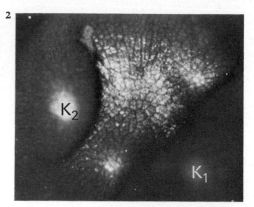

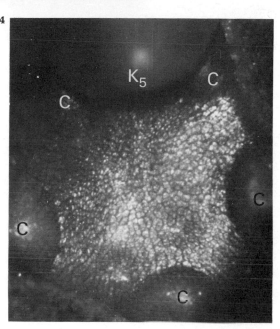

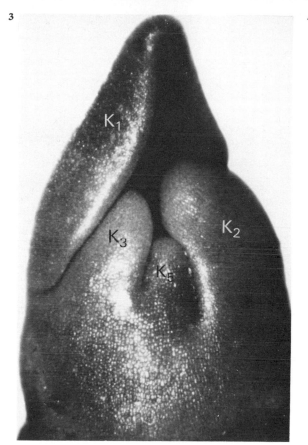

1 Top view of floral bud showing sequential inception of the five sepal primordia (K_1, K_2, K_3, K_4, K_5). x 85
2 Higher magnification of central portion of the bud shown in figure 1. x 146
3 Side view of floral bud in a developmental stage like that of figure 4. x 85
4 Top view of floral bud from which all but one sepal primordium (K_5) were removed to show the petal primordia (C) and possibly very faintly a first indication of androecial inception. x 146

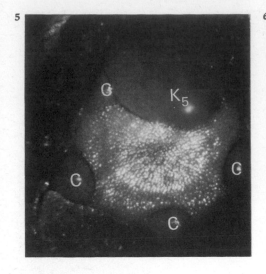

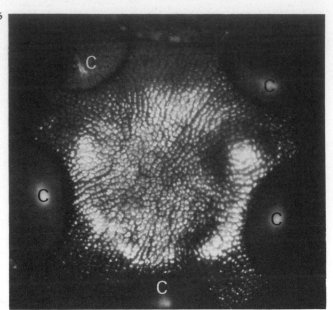

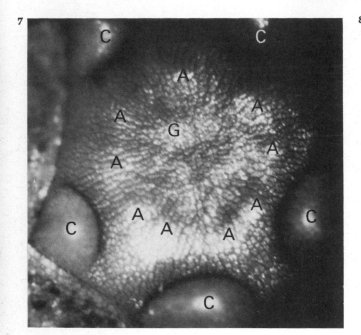

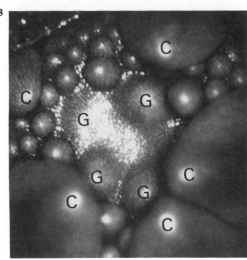

to distinguish the first sepal primordium from the leaf primordium which precedes it. The transition from the vegetative region to the floral one is gradual in this species which has solitary terminal flowers. In the mature flower, the sepals are distinct in their shape from the preceding leaves.

The five petal primordia are initiated in a very rapid sequence. In most buds one can distinguish the fourth and fifth petal primordia in terms of size. But the first three petal primordia are almost of the same size (4-8). Therefore, one cannot even be certain whether they are formed in a very rapid sequence or simultaneously. In some buds all five petal primordia are almost the same size (10). In the floral buds with distinct size differences between at least some of the petal primordia, the three large petal primordia are toward one side of the floral bud and the two smaller ones are toward the other side (7, 8). This indicates a sequence of origin which is quite different from the spiral sequence of sepal formation.

The inception of the stamen primordia is very difficult to describe. One can say with certainty that the first stamen primordia which are formed are the innermost ones and that all the other stamen primordia are formed centrifugally (6-11). One can also observe clearly that the first stamen primordia are formed in regions alternate to the petal primordia (6-9). They appear in groups of two or three, or perhaps singly, and are then followed by further primordia which appear laterally and outside of them and in alternation with each other. There seems to be considerable variation from one bud to another and even within one floral bud. Partly due to this variation and indefiniteness, several questions are difficult to answer: Are the individual stamen primordia formed on the floral apex or on common androecial humps which alternate with the petal primordia? When the individual stamen primordia become visible, they are located on inconspicuous humps which are more or less connected with each other due to growth between them. It is difficult to decide whether these inconspicuous common humps are formed before, during, or after the inception of the individual stamen primordia. Are the inconspicuous common humps initiated in a sequence similar to the one of the corolla, or are they formed simultaneously? In most floral buds they are definitely higher toward one side of the bud. This may indicate a sequence of inception as described for the corolla. As more stamen primordia are formed, the original gaps between the groups of the first stamen primordia disappear completely, and groups of stamens are no longer recognizable (10, 11, 12). As the stamens are initiated and develop, there is some upgrowth of the whole androecial region. But in the mature flower this is no longer noticeable. The number of stamens per flower was not counted. Wilson (1965) reported 150-200 stamens per flower. Each stamen consists of a filament and an introrse anther (22).

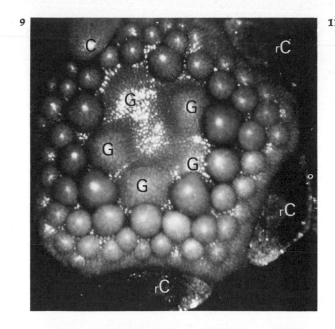

9

11

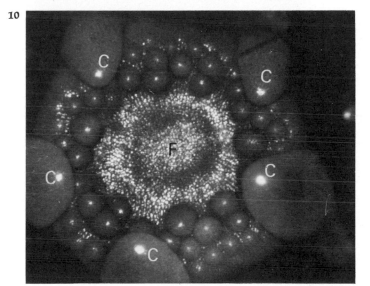

10

5 Top view of floral bud from which all but one sepal primordium (K_5) were removed. Four of the five petal primordia (C) are visible, and the androecium has an unusual almost ring-like appearance. x 85

7 Top views of floral buds from which sepal primordia were removed to show inception of stamen primordia (A) which are labelled only in figure 7. x 146

9 Top views of one floral bud before (8) and after (9) removal of four petal primordia (C). There are five gynoecial primordia (G) in the center and many stamen primordia (unlabelled). x 85

0 Top view of a floral bud of a similar developmental stage as that in figures 8 and 7, but lacking distinct gynoecial primordia. F = floral apex. Sepal primordia were removed. x 85

1 Top view of a floral bud from which young sepals and petals were removed to show young stamens and four of the pistil primordia. x 29

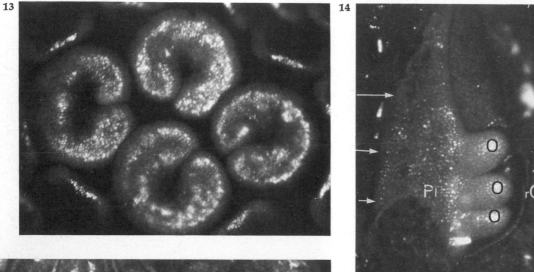

12

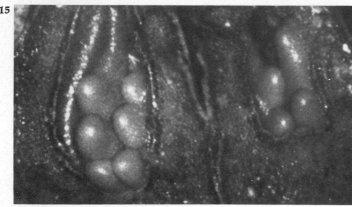

13

14

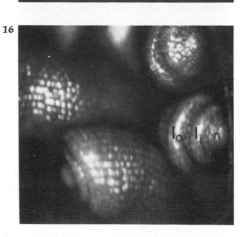

15

16

The gynoecial primordia are formed in a spiral sequence soon after the inception of the first stamen primordia. During their early development, they very much resemble stamen primordia (8, 9). But then marginal growth leads to the typical carpel shape (11-13). Postgenital fusion occurs along the margins of each pistil which develops an ovarial, stylar, and stigmatic portion. Two placentae are formed inside the margins of each pistil (i.e., the placentation may be called submarginal or even laminar). On each placenta two or three ovule primordia are initiated at about the same time (14). On some placentae one (or two) ovule primordia appear slightly smaller which may indicate a sequential origin in these cases (15, 16). Each ovule primordium forms two integuments (16-19) and later, at the base of the funiculus, an aril (20). The opening of the inner integument does not coincide with that of the outer one (21). The mature ovule is anatropous.

OTHER AUTHORS

I found no literature on the floral development of *Hibbertia scandens*. Payer (1857) described the floral organogenesis of *Hibbertia grossulariæfolia*. In contrast to my observations, he reported simultaneous origin of the five petal primordia. Furthermore, he pictured five conspicuous androecial humps (alternately to the petal primordia) on which the individual stamen primordia appear centrifugally. Instead of five carpel primordia he observed two pentamerous carpel whorls.

Recently, Wilson (1965) published a detailed investigation on the floral anatomy of the genus *Hibbertia*. He reported five or more carpels with six to eight ovules. Dickison (1968) described the carpel morphology in this genus. Both Wilson and Dickison studied only mature flowers. Older literature on the morphology of the mature *Hibbertia* flower is quoted by these authors.

BIBLIOGRAPHY

Dickison, W.C. 1968. Comparative morphological studies in Dilleniaceae, III. The carpels. J. Arnold Arb. *49*: 317-329.

Payer, J.B. 1857. *Traité d'organogénie comparée de la fleur. Texte et Atlas.* Paris: Librairie de Victor Masson.

Wilson, C.L. 1965. The floral anatomy of the Dilleniaceae I. *Hibbertia* Andr. Phytomorphology *15*: 248-274.

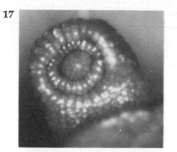

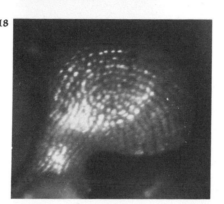

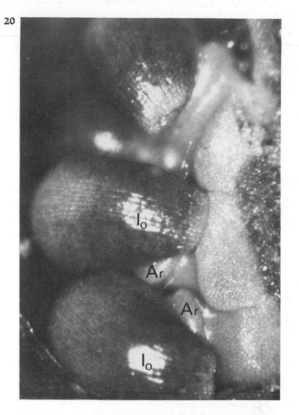

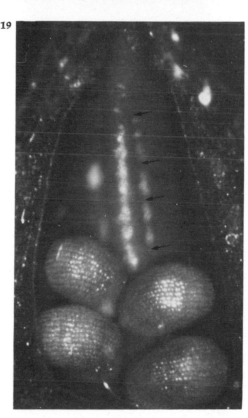

12 Top view of four young pistils and stamens at a developmental stage like that of figure 11. x 85

13 Top view of four young pistils showing formation of stigmas. x 85

14 Side view of longitudinally dissected young pistil showing one placenta (Pl) with three ovule primordia (O). The arrows point to the margin of the gynoecial appendage. rOa = ovary wall removed. x 85

15 Two adjacent young pistils from which the outer (abaxial) portions were removed to show placentae with ovule primordia. x 85

16 Three young ovules in different stages of development showing the two primordial integuments (I_o, I_1) and nucellus (N). x 146

17 Young ovule showing primordia of inner and outer integuments. x 146

18 Side view of the young ovule of figure 17. x 146

19 Young pistil whose outer (abaxial) portion was removed to show its four ovule primordia. Arrows point to the suture of the gynoecial margins which are out of focus. x 85

20 Placenta with three ovules showing young aril (Ar). x 85

21 Dissected ovule with arrows pointing to the micropyles of inner (I_1) and outer integument (I_o). x 146

22 Mature flower. x ca. 1

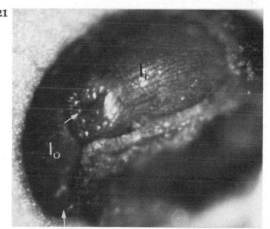

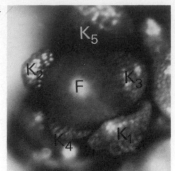

GUTTIFERAE

Hypericum perforatum L.
(common St John's-wort)

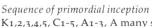

Floral diagram

Note: The contortion of the corolla is slightly visible at its base. It may be clockwise or counterclockwise.

Floral formula: ✳K5 C5 A3 × (∞) $\underline{G}$(3) O∞

Sequence of primordial inception
K1,2,3,4,5, C1-5, A1-3, A many sequentially and overlapping in origin with G1-3, O many acropetally in very rapid succession.

DESCRIPTION OF FLORAL ORGANOGENESIS

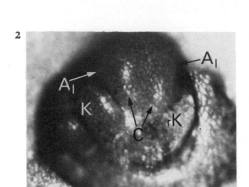

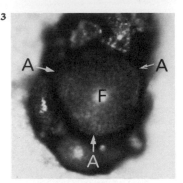

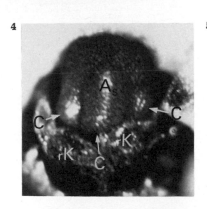

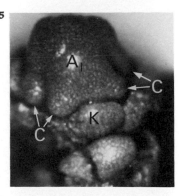

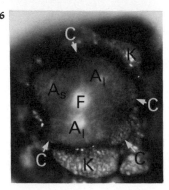

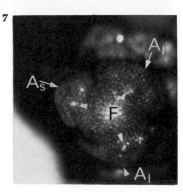

On the dome-shaped floral apex the five sepal primordia are initiated first in a distinct spiral sequence (1). Then the primordia of the five petals and the three androecial fascicles are formed at about the same time. It is possible that petal inception precedes the origin of the androecial fascicles, but if there is any time difference, it is so small that it is difficult to ascertain. Among the three fascicle primordia, two are larger in lateral direction (Al) than the third one (As) (3, 4, 5). The smaller fascicle primordium (As) is opposite one petal primordium (4). During their inception, growth occurs in the region between them. In fact, the fascicle primordium at first is indistinguishable from the subtending petal primordium. Only in a later stage of development do they become distinct due to suppression of growth in the region between them (4, 9). Each of the two large fascicle primordia (Al) is in contact with two petal primordia on its flanks due to growth between them (5, 6). During their inception, each large fascicle forms one elongate hump of tissue which is continuous with the two petal primordia on its sides (2, 5). As in the case of the small fascicle primordium and its subtending petal primordium, growth becomes suppressed in the region between the petal and fascicle primordia. And as a result of this, the larger fascicle primordia become distinct from the two petal primordia on their flanks (8). In the mature flower, the androecial fascicles are completely distinct

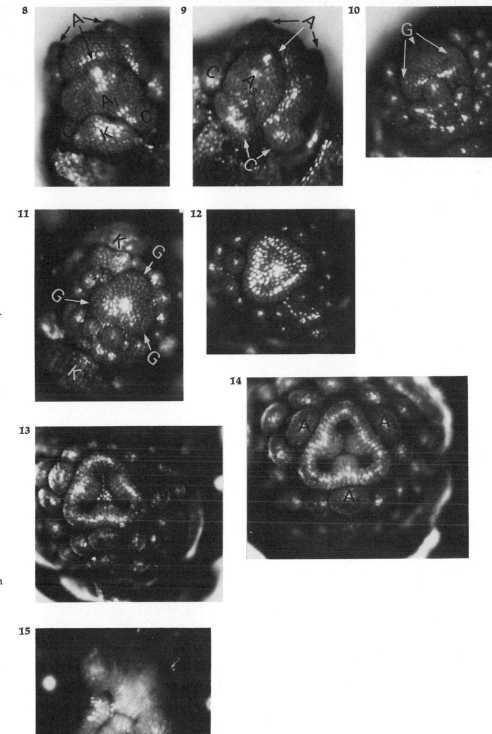

1 Top view of a floral apex (out of focus) with sepal primordia (K). x 146

2 Side view of a young floral bud from which one sepal primordium was removed (rK) to show two petal primordia (C) which are continuous with two large fascicle primordia (Al). x 146

3 Top view of floral bud in a similar stage of development as that in figure 2 showing the three fascicle primordia. Petal primordia are hardly visible. x 146

-5 Side views of floral bud slightly more advanced than the one of figure 3. In figure 4 the small fascicle primordium (As) is visible opposite one petal primordium (C). In figure 5 a large fascicle primordium (Al) with petal primordia (C) on its flanks is shown. x 146

-7 Top views of floral buds with fascicle primordia on which the first stamen primordium (see arrowhead, figure 7) is initiated. In figure 6 the focus is on the petal primordia (C) below the fascicle primordia. Sepal primordia were removed except two in figure 6. x 146

-9 Side views of floral buds showing inception of the first stamens (A) on the fascicle primordia. x 146

10 Side view of floral bud showing centrifugal stamen inception and origin of gynoecial primordia (G). x 146

-5 Top views of floral buds showing stages of androecial and gynoecial development. Origin of placentae is shown in figures 13 and 14 opposite the first (middle) stamen primordium (A) of each fascicle. x 146

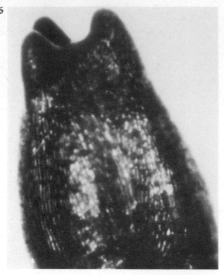

16

17

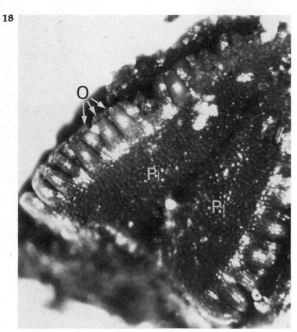

18

from the petals. Thus, the mature flower gives no indication at all that the inception of the corolla and androecium are linked spatially and temporally.

The primordia of the individual stamens appear centrifugally on the three fascicle primordia in the following fashion. First one stamen primordium is initiated in a median position on the fascicle primordium (7, 8, 9). Then two stamen primordia appear, one on each side of the first one and slightly outside it (8, 9). Additional stamen primordia are formed in alternation with the previous stamen primordia (10, 11). In this fashion up to 28 stamens may arise on one fascicle primordium. The number of stamens per fascicle is not constant. On the two larger fascicles a variation in stamen number from 24 to 28 was observed; on the smaller fascicle the number varied from 20 to 22. This variation occurred on four flowers whose stamens were counted. A larger sample probably would have yielded a greater variation.

As the stamens mature, they develop long filaments and introrse anthers. The fascicle primordium grows a little so that in the mature flower the stamens of one fascicle have a distinct common base (21).

The inception of the gynoecium occurs after the formation of the first stamen primordia, i.e. it overlaps temporally with stamen inception. Three gynoecial primordia are initiated alternately to the fascicle primordia (10, 11). Immediately after their inception, growth extends into the region between them, thus giving rise to a three-lobed triangular gynoecial ridge (11-14). As this ridge grows upward, forming the pistil wall, three placentae are initiated alternately to the gynoecial primordia (13, 14). They grow toward the center of the ovary (14). Then each placenta forms two lobes which grow outward (19). On each lobe many ovule primordia are formed in a very rapid acropetal succession (18). Each ovule develops two integuments and becomes

anatropous. The outer integument forms indentations at its rim (20). When the ovarial portion of the pistil has reached quite a considerable size, the original gynoecial primordia are still quite small (15, 16). However, in later stages of pistil development the gynoecial primordia elongate considerably, each one forming a style with a terminal stigma (17, 22). In the mature flower these styles are widely separated (22), whereas in younger stages they touch each other (17).

OTHER AUTHORS

Recently, Leins (1964) described the early floral development of *Hypericum hookerianum* and *Hypericum aegypticum*. My observations on *Hypericum perforatum* agree with Leins' description of *Hypericum aegypticum* which has the same floral diagram as *H. perforatum* disregarding the lower number of stamens per fascicle. Leins noted that the primordia of the androecial fascicles are initiated as extensions of the petal primordia. Afterwards the primordia of the petals and fascicles become distinct as reported here. Leins also noted that the inception of the petals precedes that of the androecial fascicles. As pointed out above, this may be true also for *Hypericum perforatum*. But the time interval between the inception of corolla and androecium – if it exists at all – is so short in this species that it could not be demonstrated with certainty. Payer (1857) and Sachs (1874, quoted by Leins) who studied the floral organogenesis of *Hypericum perforatum* stated that the petal primordia are initiated before those of the androecial fascicles. But Hofmeister (1868) and Hirmer (1917), both of whom are quoted by Leins, claimed that the petal primordia are formed after the primordia of the androecial fascicles.

Nelson (1954, p. 60), using mature flowers, compared the number of stamens of the large and small fascicles. The average number for the two large fascicles turned out to be 26 and 29, and that for the small fascicle 22.

Rehder (1911) observed a plant of *Hypericum nudiflorum* whose flowers showed three to ten deformed carpels (pistillodes) between the normal stamens and the pistil.

BIBLIOGRAPHY

Leins, P. 1964. Die frühe Blütenentwicklung von *Hypericum hookerianum* Wight et Arn. und *H. aegypticum* L. Ber. Deutsch. Bot. Ges. 77: 112-123.

Nelson, E. 1954. *Gesetzmäßigkeiten der Gestaltwandlung im Blütenbereich.* Chernex-Montreux: Verlag E. Nelson.

Payer, J.B. 1857. *Traité d'organogénie comparée de la fleur. Texte et Atlas.* Paris: Librairie de Victor Masson.

Rehder, A. 1911. Pistillody of stamens in *Hypericum nudiflorum*. Bot. Gaz. 51: 230-231.

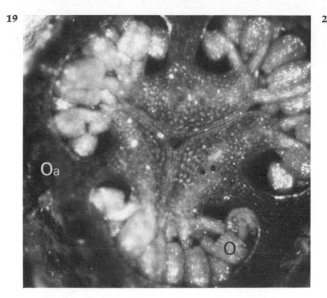

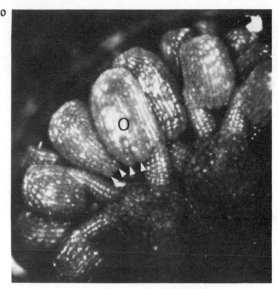

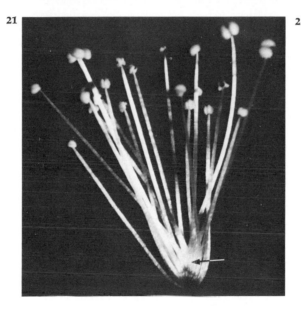

16 Side view of young pistil with small style primordia (i.e. the original gynoecial primordia) on top. x 146

17 Top view of developing styles. x 146

18 Side view of two placentae showing ovule inception. x 146

19 Cross-section through ovary showing placentation. Oa = ovary wall. O = ovule. x 85

20 Ovules with indented outer integument (arrows point to 'teeth'). x 146

21 Stamen fascicle of mature flower. Arrow points to the common base of the individual stamens. x ca. 3

22 Mature flower. x ca. 4

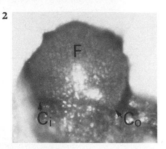

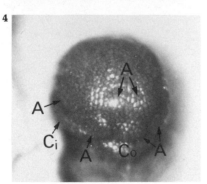

15 / PAPAVERALES (RHOEADALES, CRUCIFERALES, BRASSICALES)

PAPAVERACEAE

Chelidonium majus L. (celandine)

Floral diagram

Floral formula: + K2 C2 + 2 A∞ $\underline{G}$(2) O∞

Sequence of primordial inception
K1-2, C_o1-2, C_i1-2, A ca. 18 in partial succession G, O many in acropetal succession.
Note: For details of stamen inception see the description below. Since the circular gynoecial primordium becomes two-lobed and finally four-lobed, one could add two sets of two gynoecial primordia (i.e., $G_{circular}$, G1-2, 3-4).

DESCRIPTION OF FLORAL ORGANOGENESIS

On the dome-shaped floral apex the two sepal primordia are initiated first at about the same time (1). Then the two outer petal primordia arise simultaneously in alternation with the sepal primordia (2, 3). The inner petal primordia are initiated immediately after the outer ones. They appear also simultaneously and at a slightly higher level than the outer ones (2, 4). In contrast to the sepal primordia which soon completely overarch the floral apex, the petal

primordia grow slowly during the earlier stages of floral development (6).

Soon after the inception of the inner petal primordia, the first six stamen primordia are initiated at about the same time; four of them alternate with the four petal primordia, and two of them are opposite the outer petal primordia (4, 5). Shortly thereafter, a stamen primordium is formed opposite each of the inner petal primordia (5). Concomitantly (or perhaps even before), four stamen primordia appear at a higher level and in alternation with the stamen primordia opposite the outer petal primordia (4, 5). These are followed by four stamen primordia alternate to the stamen primordia which are opposite the inner petal primordia. Altogether, sixteen stamen primordia are formed as described above. Very roughly speaking, eight of them are arranged in a first (outer) whorl and eight of them in a second whorl alternating with the ones of the first whorl. Finally, a third whorl of stamen primordia is formed, whose members are opposite the ones of the first (outer) whorl (7). This whorl may consist also of eight primordia; but less than eight primordia were found quite frequently (8). This variation in the number of primordia in the third (inner) whorl accounts for much of the variation in stamen number of mature flowers. In about 10 flowers the number of stamens varied from 18 to 26. There may be other irregularities in the development of the androecium, especially in flowers with more than 24 stamens. However, in the buds dissected for the present study, no evidence for such other irregularities was found. Besides that, it should be noted that the stamen primordia of the inner petal quadrants are at a slightly higher level than the corresponding ones of the outer petal quadrants. Consequently, twice as many whorls, i.e. six whorls, might be described in this flower. If one considers the additional irregularities that are rather common, even

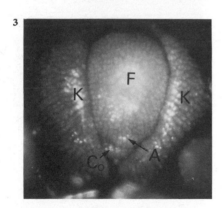

the notion of six whorls appears as an over-simplification. Therefore, it might be best to say: after (or concomitantly with) the formation of eight outer stamen primordia, further stamen primordia tend to arise in alternation with the preceding ones. Each of the stamen primordia forms a filament and an extrorse anther (9-12, 17).

During stamen inception, the distal area of the bud which will give rise to the gynoecium becomes distinct as a little dome on the androecial region below it (5, 6, 7).

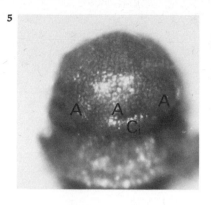

5

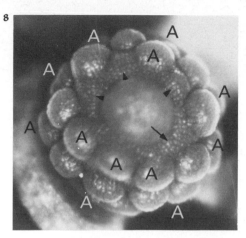

8

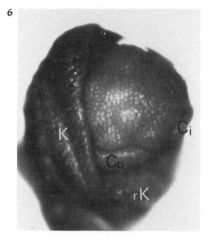

6

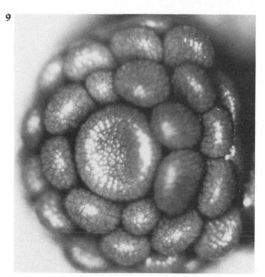

9

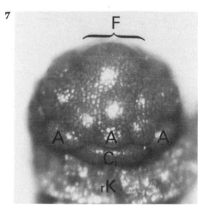

7

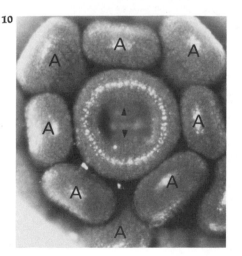

10

---

1 Side view of young inflorescence tip showing three floral buds with sepal primordia (K) and one floral apex (F) before the inception of primordia. x 146

2 Side view of floral bud from which the sepal primordia were removed to show primordia of outer (C_o) and inner (C_i) petals. x 146

3 Floral bud in about the same developmental stage as that of figures 4 and 5. x 146

4-5 Two side views of the same bud after removal of sepal primordia. Note that the stamen primordia (A) are more developed on the outer petal (C_o) quadrants than on the inner petal (C_i) quadrants where they are scarcely visible. x 146

6 Side view of a floral bud from which one sepal primordium was removed. Opposite the outer petal primordia (C_o) the stamen primordia of all three whorls are visible. x 146

7 Side view of a floral bud slightly older than that of figure 6 and viewed from an angle 90 degrees different from that of figure 6. Both sepal primordia were removed (rK). Note the distinct floral apex (F) which will form the gynoecium. Only the stamen primordia of the outer whorl are labelled. x 146

8 Top view of a floral bud from which sepal primordia were removed. Only four stamen primordia (A) of the inner whorl are well developed, three are much smaller (see arrowheads), and one is missing (see arrow). The stamen primordia of the second whorl are unlabelled, and the stamen primordia of the first (outer) whorl are accompanied by an A. x 146

9 Floral bud showing inception of the gynoecium. Sepal primordia were cut off. x 146

10 Center of floral bud showing young gynoecium with incipient placentae (arrowheads). The stamen primordia of the inner whorl are labelled (A). x 146

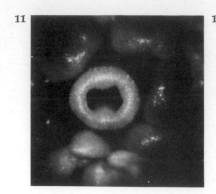

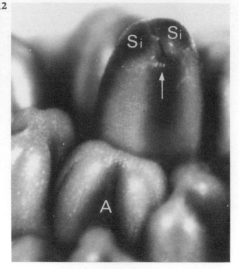

After all stamens have been initiated, a girdling gynoecial primordium is initiated on this little dome (10). During the up-growth of this girdling primordium, two placental primordia appear on opposite sides of the young cylindrical pistil (10, 11). Eventually the two regions of the gynoecial primordium which alternate with the placental primordia grow faster. Thus, the young pistil tip becomes two-lobed. The two lobes become appressed to each other, close the pistil, and form the two stigmas. Before the stigmas differentiate, two small lobes arise opposite and in continuation with the placental primordia (12). Many ovule primordia are initiated on the placentae in rapid acropetal succession (13). Each of them forms two integuments (14, 15, 16) and eventually an elaiosome.

OTHER AUTHORS

In a comprehensive developmental study of the Papaveraceae, Bersillon (1955) described the floral development of *Chelidonium majus* L. As in other genera of this family, he distinguished three early phases (étapes) of development (pp. 326-334): *a*, the formation of sepal primordia; *b*, the formation of petal primordia, a broad circular bulge (bourrelet androcéen) on which the individual stamens will form, and as a remainder the little dome in the center which will form the gynoecium; *c*, the independent development of the floral appendages. He did not describe in detail the pattern of stamen inception. However, he reproduced cross-sections through two flowers, one with 17 stamens (figure 206), and one with 24 stamens (figure 207, p. 363). In both flowers he recognized two octomerous whorls of stamens. The additional stamens he did not ascribe to a third whorl. In flowers with more than 24 stamens he could not even recognize two outer octomerous whorls. His report of gynoecial development agrees with my observations. The inception of the ovary wall is described as follows: 'le primordium (du pistil) est un bourrelet circulaire, légèrement renflé en deux points diamétralement opposés' (p. 397).

Payer (1857) also described the organogenesis of this species. Although most of his observations are confirmed here, there are some discrepancies. According to Payer, the stamens of the outer whorl are formed in pairs opposite the four petal primordia. I never saw such an arrangement of the outer stamen primordia. With regard to the third (inner) whorl Payer wrote: 'Le troisième verticille, *quand il existe,* alterne avec le second, et les étamines qui le composent naissent toutes à la fois' (italics mine). Payer's drawings on pistil development agree with my observations. But in the text, Payer claimed that the pistil wall is formed by two primordia which fuse congenitally due to interprimordial growth. He also claimed that each placenta becomes two-lobed in its full length, and that the first ovules are initiated in the middle portion of the placentae.

Schumann (1890) tended to agree with Payer's description. He noted that the placentae are formed successively, an observation which is also not supported by my findings. In addition to the normal flowers, he found flowers with three sepals.

Benecke (1880, 1882, quoted by Murbeck 1912) reported two tetramerous, one octomerous, and one hexamerous whorls of stamens which arise in succession. Possibly the two tetramerous whorls may be formed simultaneously. Although this report does not agree completely with my findings, it comes closer to them than Payer's description.

Murbeck (1912) who reproduced diagrams of cross-sections through 59 buds found that only two of those buds were identical. There is an enormous variation in the number and position of the stamens. According to Murbeck the only common

feature of all the buds is the centripetal inception of the stamens. (The interpretive formula which he proposes for the androecium is $A2^m + 2^{m-n}$.)

Recently, Merxmüller and Leins (1967) published a paper comparing the floral development of the Papaveraceae and Brassicaceae. For their developmental comparisons they used mainly *Glaucium flavum. Chelidonium majus* is briefly mentioned with respect to the mature gynoecium.

BIBLIOGRAPHY

Bersillon, G. 1955. Recherches sur les Papaveracées; contribution à l'étude du développement des Dicotylédones herbacées. Ann. Sci. Nat., Bot., ser. 11, *16*: 225-448.

Merxmüller, H. and Leins, P. 1967. Die Verwandtschaftsbeziehungen der Kreuzblütler und Mohngewächse. Bot. Jahrb. *86*: 113-129.

Murbeck, Sv. 1912. Untersuchungen über den Blütenbau der Papaveraceen. Kungl. Sv. Vet. Akad. Handlingar *50* (1): 1-168 and plates.

Payer, J.B. 1857. *Traité d'organogénie comparée de la fleur. Texte et Atlas.* Paris: Librairie de Victor Masson.

Schumann, K. 1890. *Neue Untersuchungen über den Blütenanschluss.* Leipzig.

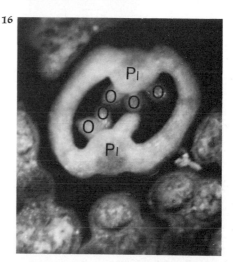

11 Top view of center of more advanced floral bud. x 85

12 Side view of young pistil showing stigma primordia (Si) and alternating smaller primordia (arrow). x 85

13 Side view of dissected young pistil showing acropetal inception of ovules. x 85

14 Side view of more advanced pistil part of whose wall was removed to show the two parietal placentae with ovule primordia. x 85

15 Portion of a placenta with ovule primordia. x 85

16 Cross-section through a young pistil showing placentae (Pl) with ovule primordia (O). x 85

17 Mature flower. x ca. 3

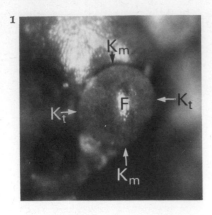

1

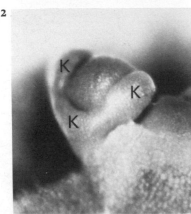

2

16 / PAPAVERALES (RHOEADALES, CRUCIFERALES, BRASSICALES)

CRUCIFERAE

Cheiranthus cheiri L. (wallflower)

Floral diagram

Floral formula: $\cdot | \cdot$ K4 C4 A2 + 4 $\underline{G}$(2) O∞
Note: Merxmüller and Leins (1967) wrote G(4) instead of G(2).

Sequence of primordial inception
K1-4, C1-4, A$_o$1-2, A$_i$1-4, G, O1-∞
Note: At least some of the sepal primordia are initiated in a very rapid succession (see text). The oval gynoecial primordium is probably formed by four gynoecial primordia which become connected by interprimordial growth. Consequently, one might write G1-4 instead of G. In later developmental stages two median lobes (gynoecial primordia) develop into stigmas which are connected by two transversal bulges.

DESCRIPTION OF FLORAL ORGANOGENESIS

The four sepal primordia are initiated at nearly the same time. The abaxial one appears first. It is immediately followed by the two transversal sepal primordia which seem to appear simultaneously. The adaxial sepal primordium seems to be formed immediately after the transversal ones (1, 4).

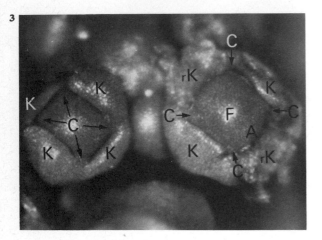

3

As the four sepal primordia are initiated, some growth occurs between them. Consequently, in the very early stages the sepal primordia are continuous with each other, i.e., they appear on a common zone surrounding the floral apex. This zone does not grow further. Therefore it is not noticeable in later stages of development. In contrast to the two median sepal primordia the two transversal sepal primordia are conspicuously curved upward with their margins. As a result, they form little pockets on the side of the floral apex (2-5). Their median portion is at a lower level than their margins.

The four petal primordia appear at the same time alternately to the sepal primordia (3). They are much smaller than the sepal primordia and occupy only a very small area of the floral bud (3-6, 9-10). They grow very slowly in comparison to all other floral organs. In the late stages of floral development they reach their final size and shape relatively fast (17).

After petal inception, the two outer stamen primordia are initiated at the same time (3). They are situated between the petal primordia at about the same level as the latter. This is spatially possible because of the special curved shape of the transversal sepal primordia. The four inner stamen primordia appear as two pairs at a much higher level than the primordia of the petals and outer stamens (4-6). Each pair of them is opposite one median sepal primordium. There is no upgrowth between the primordia of each pair. The primordia are initiated separately and remain distinct during their further development. Perhaps a slight amount of interprimordial growth may occur during later stages of development. Each of the inner stamen primordia as well as the outer ones forms a filament and an introrse anther.

After stamen inception, a gynoecial ridge is formed on the floral apex. Due to the greater dimension of the floral apex in the transversal plane, the gynoecial ridge is elliptical (8-9). As this ridge grows upward, it forms a pistil wall whose major portion constitutes the ovary wall and whose minor upper portion differentiates into two stigmatic lobes. The stigmatic lobes are in the median plane. They are continuous with two bulges in the transversal plane (14). The formation of the two conspicuous stigmatic lobes and the distinct lateral lobes occurs gradually (10-12, 14). However, it seems to be visible already on the incipient gynoecial ridge (8-9). Hence,

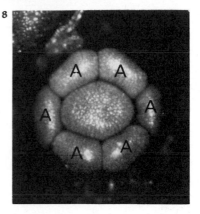

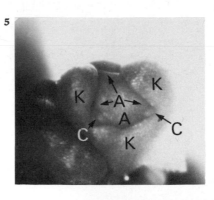

---

1 Young floral bud showing the median (K_m) and transversal sepal primordia (K_t) and growth between the four sepal primordia. F = floral apex. x 146

2 Side view of a floral bud showing the two median and one of the lateral sepal primordia, the latter of which forms a pocket. x 146

3 Top view of two floral buds showing petal inception (C). In the bud to the right the two median sepal primordia were removed (rK); the two outer stamen primordia (A) are faintly visible in this bud. x 146

4 Side view of a floral bud showing one of the outer stamen primordia (A) between two petal primordia (C); the two pairs of inner stamen primordia (see arrows) are just being initiated. Also a younger floral bud like that in figure 1 can be seen. One of the transversal sepal primordia of this bud is hidden. x 146

5 Side view of a floral bud with the primordia of the calyx, corolla, and androecium. Only three of the four inner stamen primordia are visible. x 146

6 Top view of a floral bud from which the young calyx was removed to show primordia of petals and stamens. x 146

7 Top view of unusual bud with five stamen primordia (A) of unequal size and shape. Young calyx was removed. x 146

8 Top view of floral bud from which sepal primordia were removed to show stamen primordia (A) and incipient gynoecium. x 146

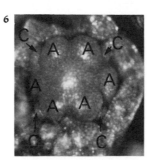

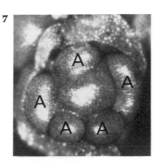

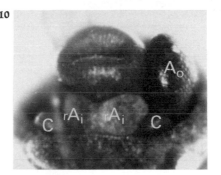

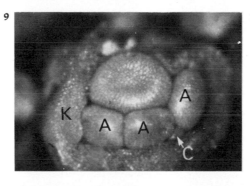

9 Dissected floral buds showing stages in gynoecial development. In figure 10 the pair of inner stamen primordia was removed (rA$_i$); one of the outer stamen primordia is still present (A$_o$). x 146

12 Top views of gynoecia showing incipient placentae. x 146

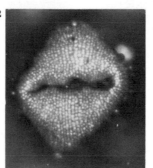

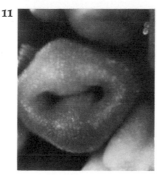

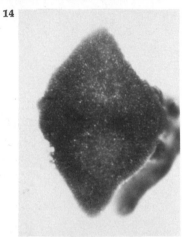

one might even say that the gynoecium originates as four primordia which become instantly connected by interprimordial growth.

During the upgrowth of the pistil wall, two pairs of placental bands are formed in the same plane as the median stigmatic lobes (12, 13). Immediately after their inception many ovules are initiated on these bands. Consequently four rows of ovules are formed (15, 16). Each ovule becomes bitegmic and anatropous. Immediately after the inception of the placental bands, the region between the bands of each pair grows out thus forming a septum. The two septa meet in the center of the pistil and fuse postgenitally (15). An upward growth occurs below the two septa so that they become inserted on a common septum which is continuous with them and the ovary wall. The common septum is not indicated in the floral diagram.

A two-lobed gland develops late on the adaxial base of each outer stamen.

A number of floral buds with an unusual number and position of appendages were found. For example, in one bud only a single inner stamen primordium occurred instead of one of the stamen pairs (7). Consequently, the bud had only five stamen primordia. The outer ones were broader than in the ordinary floral bud.

OTHER AUTHORS

A voluminous and controversial literature deals with the mature structure, development, and interpretation of cruciferous flowers (see Eichler 1865; Eames and Wilson 1928; Eggers 1935; Puri 1941; Alexander 1952). To begin with the calyx, it is not surprising that much disagreement exists with regard to the inception of the sepals. They arise in such a rapid sequence (or some of them perhaps even simultaneously) that it is extremely difficult or impossible to list the sequential order of

origin. Payer (1857) assumed that in *Cheiranthus cheiri* the abaxial sepal appears first, then the two lateral ones, and finally the adaxial one. Schumann (1890, quoted by Alexander 1952) came to the same conclusion in other species of the Cruciferae. Alexander (1952), who presented microtone sections through young buds of *Cheiranthus cheiri*, claimed that the two median sepals originate before the two lateral ones. Eichler (1865) came to the same conclusion for *Erucaria allepica*, but his crucial drawing (tab. VI, figure 3) shows no difference in size between the two lateral sepal primordia and the adaxial one. This seems to indicate that at least the lateral and adaxial sepals originate at about the same time (see my figures 1, 4). Merxmüller and Leins (1967) showed by a median section through a very young floral bud of *Sisymbrium strictissimum* that the abaxial sepal primordium is initiated immediately after the adaxial one (table 6, figure 16). Finally Eggers (1935) thought that the lateral sepals are initiated before the median ones.

There is general agreement that the four petals are formed simultaneously. But opinions are again divided with regard to the androecium. Payer (1857), Eichler (1865), and others thought that each of the inner stamen pairs originates from one primordium. Merxmüller and Leins (1967) concluded as I did that all inner stamens originate as separate primordia on the floral apex. Payer (1857) claimed that the primordia of the outer stamens and inner stamen pairs follow the sequential pattern that he described for calyx inception, i.e., proceeding from the abaxial toward the adaxial side of the floral apex. I could find no evidence for this postulate.

Much controversy also exists with regard to gynoecial development. Payer (1857) noted that the gynoecium originates by two primordia in the transversal plane which become united by interprimordial growth immediately after their inception. However, he did not support this claim by a drawing. His drawings rather agree with my photographs. Eichler (1865) and others supported Payer's view of two carpel primordia. Eggers (1935) and Alexander (1952) noted that the gynoecium originates as an oval ring. The primordia of the commissural stigmas develop later on this girdling gynoecial primordium. Merxmüller and Leins (1967) demonstrated by serial cross-sections that in *Sisymbrium strictissimum* the gynoecium originates from two median gynoecial primordia. In addition to these primordia, which were called fertile carpel primordia, two sterile carpel primordia were postulated. Evidence for the latter was not shown except by interpretation of the vascularization.

Schweidler's (1911) description of the glands is in agreement with my observations. Roth (1957-58) presented a detailed account of ovule development in *Capsella bursa-pastoris*.

BIBLIOGRAPHY

Alexander, I. 1952. Entwicklungsstudien am Blüten von Cruciferen und Papaveraceen. Plant 40: 125-144.

Eames, A.J. and Wilson, C.L. 1928. Carpel morphology in the Cruciferae. Amer. J. Bot. 15: 251-271.

Eggers, O. 1935. Über die morphologische Bedeutung des Leitbündelverlaufes in den Blüten der Rhoeadalen und über das Diagramm der Cruciferen und Capparidaceen. Planta 24: 14-58.

Eichler, A.W. 1865. Über den Blütenbau der Fumariaceen, Cruciferen und einiger Capparideen. Flora 48: 433-558.

Merxmüller, H. and Leins, P. 1967. Die Verwandtschaftsbezeihungen der Kreuzblütler und Mohngewächse. Bot. Jb. 86: 113-129.

Payer, J.B. 1857. *Traité d'organogénie comparée de la fleur. Texte et Atlas.* Paris: Librairie de Victor Masson.

Puri, V. 1941. Studies in floral anatomy. I. Gynoecium constitution in the Cruciferae. Proc. Ind. Acad. Sci. 14: 166-187.

Roth, I. 1957-58. Die Histogenese der Integumente von *Capsella bursa-pastoris* und ihre morphologische Deutung. Flora 145: 212-235.

Schweidler, H. 1911. Über den Grundtypus und die systematische Bedeutung der Cruciferen-Nektarien. I. Beih. Bot. Centralbl. 27(1): 337-371.

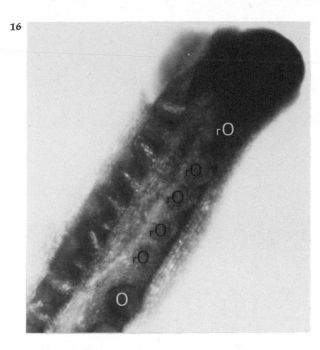

13 Internal view of the gynoecium of figure 12 after longitudinal splitting. Arrowheads point to the two bands of a placenta on which ovule primordia may just be initiated. x 146

14 Top view of young stigmas. x 85

15 Cross-section through young pistil with two septa fusing postgenitally in the center (see arrow) and ovule primordia (O) at their base. x 146

16 Longitudinally dissected young pistil showing ovule primordia in a row along one side of a developing septum. To the right all but one of the ovule primordia were removed (rO). x 146

17 Mature flower. x ca. 4

17 / ROSALES

ROSACEAE

Fragaria vesca L. (strawberry)

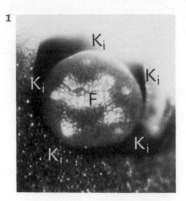

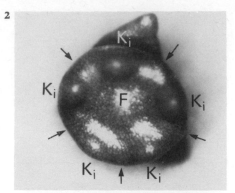

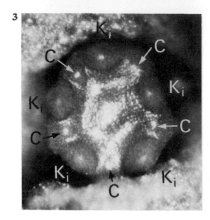

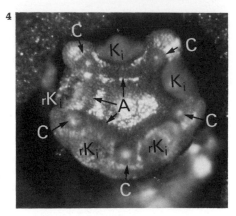

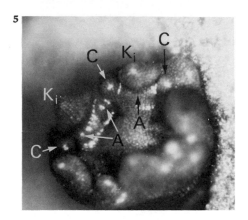

Floral diagram
Note: The number of pistils is much larger than indicated in this diagram.

Floral formula: $* \; [K_0 5 K_i 5 \; C5 \; A5 \times 3+5] \; G\infty \; O\infty$

Sequence of primordial inception
$K_i 1,2,3,4,5$, $K_0 1,2,3,4,5$, $C1,2,3,4,5$, $A_{ki} 1,2,3,4,5$ each forming three stamen primordia, $A_c 1,2,3,4,5$, $G1,2,3,4,5$ each forming first two (or one) and then centripetally many gynoecial primordia (pistil primordia) with one ovule each.
Note: The five primordia in each of the six whorls are initiated in a very rapid sequence which at least in some buds approaches simultaneity.

DESCRIPTION OF FLORAL ORGANOGENESIS

On the slightly convex floral apex the five inner sepal primordia are initiated in very rapid succession (1, 2) which at least in some buds (e.g. 1) approaches simultaneity. The sequence of inception does not seem to be a spiral one, since two smaller primordia can be seen on one side and three slightly larger primordia on the other side of the floral apex (2). These slight size differences of the sepal primordia seem to indicate that the inception of the inner calyx commences on one side of the floral apex and proceeds gradually to its other side. However, the more common spiral sequence of inception may occur at least in some of the floral buds (3). Immediately after the inception of the inner sepal primordia, the primordia of the outer calyx (epicalyx) are formed in the same sequence as those of the inner calyx. Whereas the primordia of the inner sepals grow in a vertical direction, those of the outer sepals grow out horizontally at first, and then grow upward vertically. As the primordia develop, growth occurs in a ring-like zone at the base of inner and outer calyx (2). This leads to the formation of a cup at the base of the original primordia of inner and outer sepals.

Shortly after the inception of this cup, the five petal primordia are initiated. They are situated near the upper margin of the developing cup, or in other words at the base of the outer sepal primordia (3, 4). The sequence of inception is the same as that of the sepal primordia. The shape of

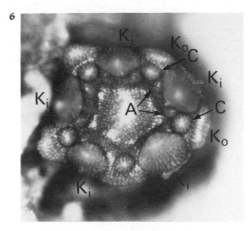

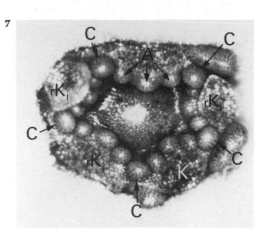

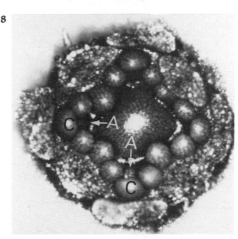

FIGURES 1-10. Top views of floral buds showing inception of inner sepals (K_i), outer sepals (K_o), androecial primordia (A) and pistils (G). In some of the photographs light reflections create the wrong impression of a pentagonal primordium inside the androecium (e.g. in figure 6). x 146

1. Inception of inner sepals.
2. Inception of outer sepals (arrows).
3. Inception of petals.
4. Inception of androecial primordia opposite primordia of inner sepals. Slightly oblique view. Three of the inner sepal primordia were removed (rK_i).
5. Slightly oblique view of a bud. The androecial primordium to the left has formed two lateral stamen primordia (A).
6. Four of the androecial primordia have formed lateral stamen primordia, two of which are labelled (A).
7. Inception of middle stamen primordium on the androecial primordium opposite the inner sepal primordium. Five groups of three stamen primordia are formed in this way, one of which is labelled. All inner sepal primordia were removed (rK_i).
8. Tetramerous floral bud from which the young calyx was removed to show inception of antepetalous stamen primordia (A).
9. More advanced floral bud; young calyx removed.

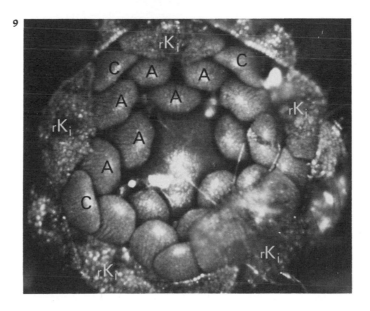

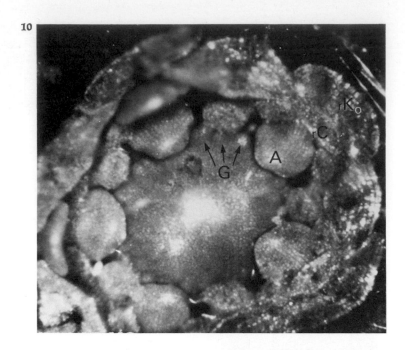

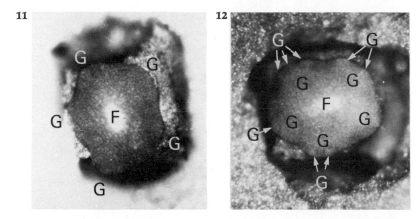

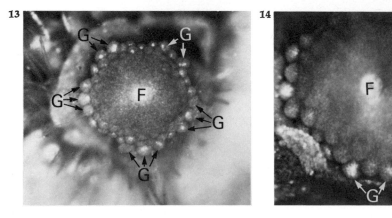

the young petal primordia is identical with the shape of the young stamen primordia. As they develop, marginal growth occurs, which finally produces the leaf-like shape of the mature petals (9, 23).

After petal inception, five laterally elongated androecial primordia are formed at the base of the inner sepal primordia on the developing cup (4, 5). These elongate primordia appear in the same sequence as the sepal primordia. Immediately after the inception of such an elongate primordium, it grows faster on the sides that face adjacent petal primordia. This accelerated growth on its sides leads to the formation of two stamen primordia (5, 6). A third stamen primordium is formed from the middle portion of the elongate primordium (7). Since the elongate androecial primordium is slightly curved with its sides away from the floral axis, the middle stamen primordium is located nearer the base of the cup than the two lateral stamen primordia (7, 8). After the inception of the five groups of three stamen primordia (i.e. 15 stamen primordia), five single stamen primordia are formed opposite the petal primordia on the developing cup (8). In radial direction these five stamen primordia are located outside the middle stamen primordia of the staminal groups and inside the lateral primordia of the staminal groups. The middle primordia of the staminal groups grow fastest of all of the stamen primordia. Also the sectors of the inner sepals grow more toward the floral apex than the petal sectors. As a result of this differential growth, the inner outline of the developing cup and its appendages becomes clearly pentagonal (7); the petal regions form the corners of the pentagon, i.e. the primordia of the petals and the stamens opposite the petals are farther removed from the floral apex than the five stamen primordia opposite the inner sepal primordia (8, 9).

After the inception of the antepetalous stamen primordia, the floral apex assumes a pentagonal form corresponding with the pentagonal inner outline of the cup. In other words, five inconspicuous primordia are formed on the flanks of the floral apex along the petal radii (11). These primordia, which may be called gynoecial primordia, are initiated in the same sequence as the sepal primordia. In their order of appearance, each of the five gynoecial primordia forms first two gynoecial primordia (pistil primordia) at its base (12). Subsequently a third gynoecial primordium (pistil primordium) may be initiated between and slightly above the first two pistil primordia (12-14). In some cases one pistil primordium seems to appear first and is followed by two pistil primordia on its flanks and on a slightly higher level (12). Both patterns occur on the same bud (12). After the inception of the five groups of pistil primordia, further pistil primordia are formed between these groups. Gradually more and more pistil primordia are formed alternating with the previous ones and in a centripetal sequence (10, 14-16). While

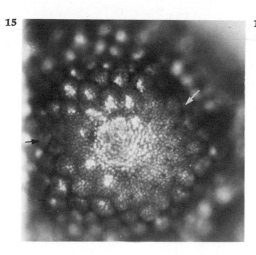

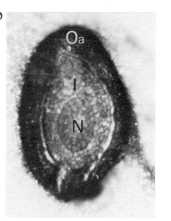

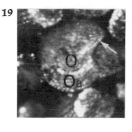

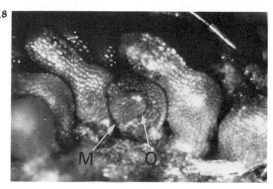

0 Older floral bud from which all organs except antepetalous stamens and gynoecium were removed.

FIGURES 11-14. Stages in gynoecial development.
1 Spiral inception of gynoecial primordia. x 146
2 Inception of gynoecial (pistil) primordia (G) on the five gynoecial primordia (G). x 146
3 Formation of pistil primordia. The first formed pistil primordia are labelled (G). x 85
4 Formation of pistil primordia in a tetramerous floral bud. x 146
6 Center of young gynoecium showing inception and early development of pistil primordia. In figure 15 arrows point to antepetalous regions of retarded pistil inception and development. x 146
7 Adaxial view of young pistil. x 146
8 Side view of young pistils one of which was dissected to show insertion of the ovule primordium (O) near margin (M) of young ovary. x 146
9 Cross-section through young pistil showing ovule primordium inserted near one margin (see arrow). x 146
0 Cross-section of older pistil and ovule showing integument (I) and nucellus (N). x 146

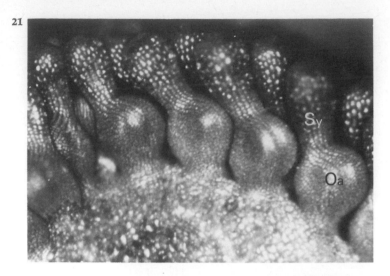

21

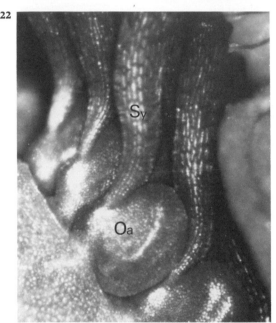

22

23

this inception of pistil primordia occurs, the floral axis continues to grow, thus providing the space for the large number of pistil primordia.

The development of the individual pistil primordia is as follows. First a primordium is formed which resembles a stamen primordium in its shape (15). Then marginal growth towards the adaxial side occurs. This leads to the formation of a hooded structure (16). Continued growth of the margins leads to the closure of the developing pistil (17). At the same time much growth of the lower outer portion of the pistil wall produces the ovary. As a result of this extreme 'bulging out' of the ovary, the style has a peculiar attachment near the base of the mature pistil (21, 22). One ovule primordium is initiated near one margin of the ovary wall (18, 19). It forms one integument (20).

In summary, one can conclude that there is much more regularity in the development of this flower than one would expect from studying the mature flower. Both androecium and gynoecium fit into the pentamerous pattern set by the calyx and corolla. There are two pentamerous whorls of androecial primordia. The antesepalous one forms 15 stamens, because each of the androecial primordia develops three stamens. The gynoecium is initiated by five inconspicuous gynoecial primordia, on which the large number of pistil primordia (secondary gynoecial primordia) is formed. All successive whorls of the developing flower are alternating with each other, with the exception of the gynoecial whorl which is opposite the antepetalous stamen primordia. Occasionally tetramerous flowers and floral buds were found (8, 14).

OTHER AUTHORS

Rauh and Reznik (1951) described briefly the floral development of this species. They concentrated their attention, however, on developmental changes within the floral axis. With regard to the androecium they mentioned that a third whorl of stamens may be present or absent (depending on the nutritional status of the plant) and the number of stamens in the second whorl varies between five and ten. They did not indicate that the floral apex becomes pentagonal (i.e. develops five primary gynoecial primordia) before the inception of the pistils.

Bessey (1898) noted that the ovules 'have their origin on the edge of one or the other lamina' of the carpel (p. 308). Schaeppi and Steindl (1950) came to the same conclusion. In rare cases they found an ovule on both sides of one carpel. Peltation is only slight according to the latter authors. Bugnon's (1929) morphological study of the epicalyx concerns mainly interpretive questions (he concluded that the members of the epicalyx are homologous to bracteoles). Jahn and Dana (1970) described the development of the inflorescence in *Fragaria ananassa*. The few comments which they added on floral development are in agreement with my observations. Haskell and Williams (1954) reported that 'sepal and petal numbers hardly varied within ploidy group, whereas stamen numbers of octoploids and decaploids were considerably higher than in lower ploidy groups' (p. 629). They did not study development. Murbeck (1915) investigated the morphological variation in 1600 mature flowers of *Comarum palustre* L., a species that has the same floral diagram as the strawberry.

BIBLIOGRAPHY

Bessey, E.A. 1898. The comparative morphology of the pistils of the Ranunculaceae, Alismaceae, and Rosaceae. Bot. Gaz. 26: 297-314.

Bugnon, P. 1929. Calicule des Rosacées et concrescence congénitale. Bull. Soc. scient. Bretagne 6: 9-20.

Haskell, G. and Hedley Williams. 1954. Biometrical variation in flowers of a polyploid series of strawberries. J. Genetics 52: 620-630.

Jahn, O.L. and Dana, M.N. 1970. Crown and inflorescence development in the strawberry *Fragaria ananassa*. Amer. J. Bot. 57: 605-612.

Murbeck, S. 1915. Über die Baumechanik bei Änderungen im Zahlenverhältnis der Blüte. Lunds Univ. Arsskrift N.F. Afd. 2, 11: 1-36.

Rauh, W. und Reznik, H. 1951. Histogenetische Untersuchungen an Blüten- und Infloreszenzachsen. 1. Teil. Die Histogenese becherförmiger Blüten- und Infloreszenzachsen, sowie der Blütenachsen einiger Rosoideen. Sitzber. Heidelberg. Akad. Wiss., Math.-naturwiss. Kl. Abh. 3.

Schaeppi, H. und Steindl, F. 1950. Vergleichend-morphologische Untersuchungen am Gynoeceum der Rosoideen. Ber. Schweiz. Bot. Ges. 60: 15-50.

22 Side views of developing pistils showing formation of ovary (Oa) and style (Sy). Figure 21, x 146. Figure 22, x 85
23 Mature flower. x ca. 2.5

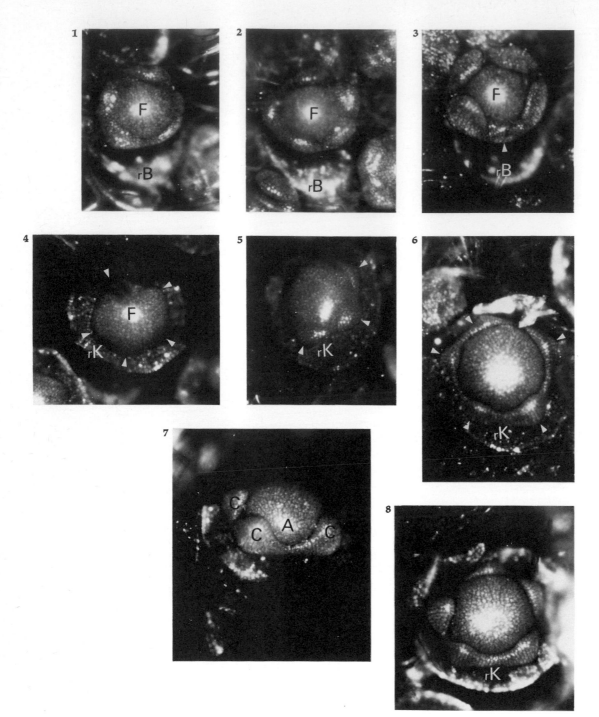

18 / ROSALES

FABACEAE (LEGUMINOSAE)

Albizia lophanta (Willd.) Bentham
(= **A. distachya** Macb.)
(plume albizia)

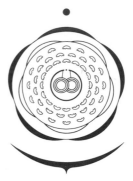

Floral diagram
Note: The inconspicuous androecial tube is not indicated.

Floral formula: $*$ K(5) C(5) A(∞) $\underline{G}$1 O10

Sequence of primordial inception
K1,2,3,4,5, C1-5, A1-5, A many in partial succession,
G1, O1-10
Note: Sepal inception may be simultaneous or close to simultaneous in some cases.

DESCRIPTION OF FLORAL ORGANOGENESIS

The sepal primordia are initiated in a very rapid succession. For this reason it is difficult to determine unequivocally the exact sequence of inception. In many floral buds the fifth sepal primordium is distinguishable (using its relative size as a criterion), whereas the sequence of the first four sepal primordia is difficult to establish, since they hardly differ in size (1, 3). In some floral buds all sepal primordia show more or less the same size (2). Extension of growth to the areas between the sepal primordia occurs before the inception of the corolla.

Thus, the calyx tube is initiated in a very early stage of development.

The five petal primordia are formed at about the same time (4, 5). As in the case of the calyx, growth soon extends to the areas between the primordia (6, 7). Thus, a corolla tube is initiated before the inception of the androecium. Relatively more growth may occur between the two petal primordia on the abaxial side and between the three other petal primordia (8). In other words, the corolla may become slightly zygomorphic. In some of the floral buds two petal primordia are more closely associated with each other than the other ones (9).

Before the inception of the individual stamens, the floral apex starts to enlarge on its flanks directly above the insertion of the developing corolla. More growth occurs in the areas alternating with the petal primordia than in the areas opposite the petal primordia (6, 7, 8). In other words, the enlarging flanks of the floral apex form five inconspicuous primordia (like bulges) alternately to the petal primordia. On these five inconspicuous primordia and the areas between them the individual stamens are initiated in very rapid acropetal succession (9, 10, 11). The size differences between the outermost and innermost stamen primordia is minimal, but it can just be noticed (11, 12). During the further development each stamen primordium forms a dorsifixed anther with four pollen sacs. The five inconspicuous androecial primordia and the areas between them continue to grow, thus forming an inconspicuous androecial tube around the base of the pistil (15).

After the inception of the stamen primordia, a crescent-shaped gynoecial primordium is formed (12, 13). As it develops its margins come into contact and fuse postgenitally (17, 18, 15). A number of ovule primordia are initiated near the fused margins inside the ovarial cavity (16, 19). Each ovule primordium forms two integuments (19). While the ovules develop, a long style becomes distinct from the ovary. It forms a stigmatic surface on its tip. One unusual floral bud had two gynoecial primordia at slightly different stages of development (14).

OTHER AUTHORS

I found no detailed literature on the floral development of *Albizia*. Hirmer (1918) reported for *Albizia lophanta* that the first five stamen primordia arise in an antesepalous position. Then, the inception of five to six further stamen primordia proceeds toward a position opposite the center of each petal primordium. In this fashion an outer whorl of $5 + 5 \times (5\text{-}6)$, i.e. 25-30, stamen primordia is formed. Subsequently four

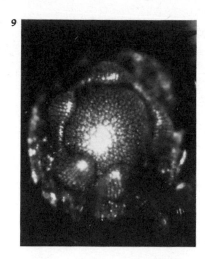

9

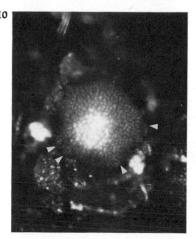

10

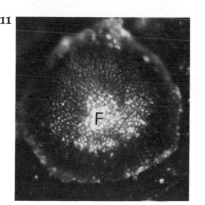

11

FIGURES 1-3. Floral buds with sepal primordia only. x 150

1 The smallest sepal primordium is to the right towards the transversal plane.
2 All sepal primordia have about the same size.
3 The smallest sepal primordium is abaxial (see white arrowhead).
-5 Floral buds with developing calyx removed to disclose the incipient petal primordia (see white arrowheads). In figure 5 only three of the five petal primordia are visible. x 150

FIGURES 6-10. Floral buds with the developing calyx removed to show the petal primordia with a developing corolla tube at their base and androecial primordia, on which the primordia of the individual stamens will be formed. x 150.

6 White arrowheads point to the petal primordia.
7 One incipient androecial primordium is visible between two petal primordia.
8 More growth occurred between the two petal primordia towards the abaxial side.
10 Two primordia of individual stamens are visible (see white arrowheads). In some of the other figures the first individual stamen primordia are faintly recognizable on the five androecial primordia which are alternating with the petal primordia.
11 Floral bud with the developing calyx and corolla removed to disclose the centripetal inception of the many stamen primordia. x 150

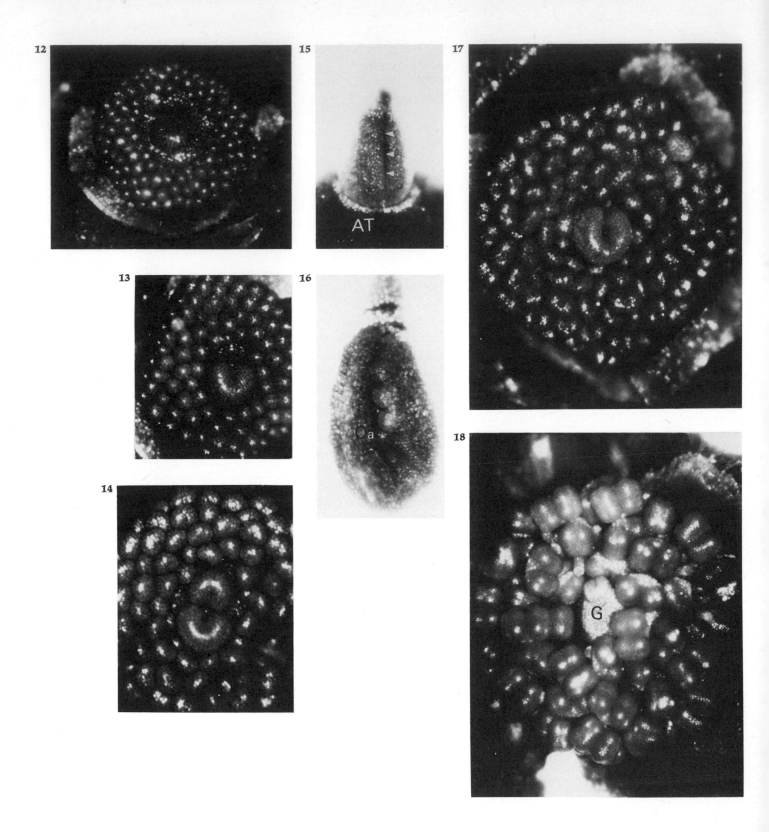

other whorls of stamens arise centripetally (p. 182).

Leinfellner (1970) describes the mature carpel by a series of cross-sections. He concludes that the lamina is epeltate and that the short stalk of the carpel is unifacial.

Newman (1936) described the floral development of two species of *Acacia*, a genus of the same subfamily as *Albizia* and with a similar floral construction. He mentioned that after petal inception, 'the outer part of the convex end of the axis (floral apex) develops into a "shoulder" .... On the "shoulder" the stamen primordia arise, beginning at its edge and in front of the sepals' (p. 58). He did not indicate whether the outline of the 'shoulder' is pentagonal (as in *Albizia*). He also did not report an androecial tube. Otherwise, there is agreement between Newman's results on *Acacia* and my observations on *Albizia distachya*.

BIBLIOGRAPHY

Hirmer, M. 1918. Beiträge zur Morphologie der polyandrischen Blüten. Flora *110*: 140-192.

Leinfellner, W. 1970. Zur Kenntnis der Karpelle der Leguminosae. 2. Caesalpiniaceae und Mimosaceae. Österr. bot. Zschr. *118*: 108-120.

Newman, I.V. 1936. Studies in the Australian *Acacias*. VI. The meristematic activity of the floral apex of *Acacia longifolia* and *Acacia suaveolens* as a histogenetic study of the ontogeny of the carpel. Proc. Linn. Soc. New S. Wales *61*: 56-88.

19

20

FIGURES 12-14. Developing androecium and gynoecium of floral buds. x 150

2 The inception of the gynoecial primordium is shown.

4 An unusual bud with two gynoecial primordia.

5 Gynoecial primordium showing postgenital fusion of its margins (see white arrowheads), and the beginning of style formation. A small androecial tube (AT) surrounds the base of the gynoecial primordium. Stamens were removed. x 83

6 Ovary with ovule primordia. The white arrowhead points to the suture of the fused margins. About one-third of the ovary was cut off to expose the ovule primordia. x 150

8 Developing androecium and gynoecium of floral buds. Figure 17, x 150. Figure 18, x 83

9 Dissected ovary with nearly mature ovules. On the upper ovule the two integuments are visible. Only the five ovules of one gynoecial margin can be seen. x 83

10 Opening flower showing the calyx, corolla, and stamens with dorsifixed anthers. At anthesis the stamens are about four times longer than the corolla. x 8

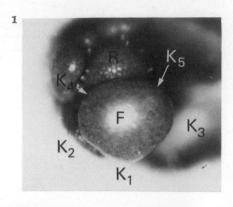

1

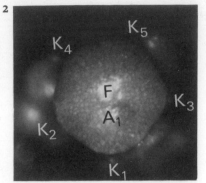

2

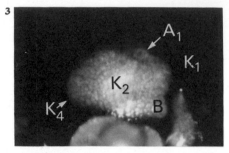

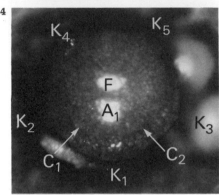

3

4

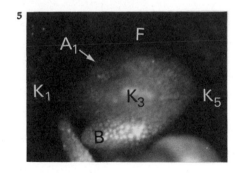

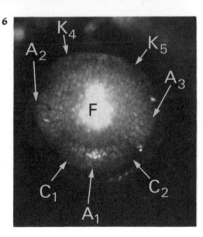

5

6

19 / ROSALES

FABACEAE
(LEGUMINOSAE)

Pisum sativum L. var. Alaska
(garden pea)

Floral diagram
Note: Petals 1 and 2 are postgenitally fused to form the keel, petal 3 forms the standard, and petals 4 and 5 constitute the wings of this papilionaceous flower (see figure 22).

Floral formula: $\cdot|\cdot$ K(5) C3{2} A(9) 1 $\underline{G}$1 Oca.6-8

Sequence of primordial inception
K1,2-3,4-5 A1, C1-2 A2-3 C3 A4 G C4-5, A5-10
Note: Some of the primordia with simultaneous origin according to this formula might be initiated sequentially as indicated in the text below. C3 and A4 might originate from one common primordium.

DESCRIPTION OF FLORAL ORGANOGENESIS

The floral apex is formed in the axil of a primordial bract on a rudimentary side axis. The first sepal primordium (K1) is initiated abaxially (1). Then a sepal primordium appears simultaneously on each side of the first sepal primordium (K2 and K3 in figure 1). Finally the last two sepal primordia (K4 and K5) are formed also at the same time (1, 2). Thus, the inception of sepal primordia begins at the abaxial side of the floral apex and proceeds rapidly to its other side. As a result of this, zygomorphy becomes slightly evident already during the inception of the calyx. Due to growth between the sepal primordia, a calyx tube is initiated immediately after the inception of the sepal primordia (1, 2). When the last two sepal primordia appear, the first stamen primordium (A1) becomes visible opposite the first sepal primordium (2, 3). Immediately afterward, the first two petal primordia (C1 and C2), which will form the keel, are initiated on the flanks of the first stamen primordium (4, 6). A little growth occurs at the base of the first stamen primordium (A1) and the first two petal primordia (C1 and C2) which creates a common basis for these three primordia (6, 8, 11). In later

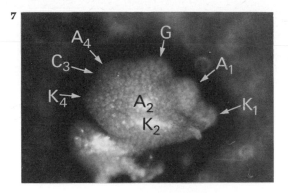

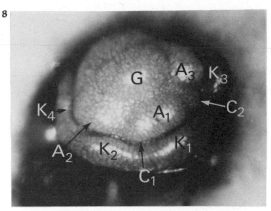

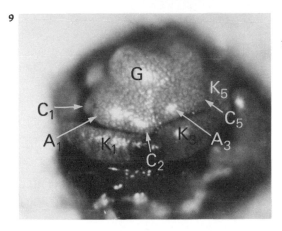

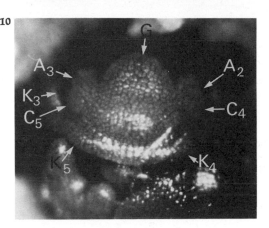

FIGURES 1–14. Top and side views of floral buds in various stages of development. Only in figures 13 and 14 were sepal primordia removed. × 146

1 Top view of floral apex (F) with the sepal primordia (K1, K2, K3, K4, K5). R = apex of rudimentary side axis.
2 Top view of floral apex with sepal primordia and first stamen primordium (A1).
3 Side view of a floral bud in the developmental stage of that in figure 2.
4 Top view of floral bud showing the primordia of the sepals, the first stamen (A1), the keel petals (C1 and C2). Other primordia may be faintly recognizable.
5 Side view of a similar developmental stage as that of figure 4.
6 Top view of floral bud showing inception of A2 and 3 in addition to A1 and the keel primordia (C1 and 2).
7 Side view of a stage slightly older than that of figure 6. The primordia of standard (C3) and the superposed stamen (A4) are not yet clearly distinguishable from each other. The gynoecial primordium may have been initiated (G).
10 Three different views of the same floral bud.
11 Top view of floral bud showing the primordia of all organs except two antepetalous stamens (A9 and 10).
12 Top view of floral bud with the primordia of all organs. The stamen primordia are not labelled.

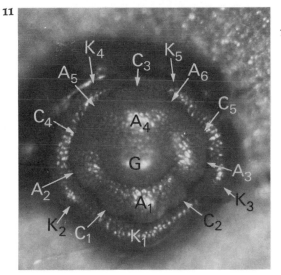

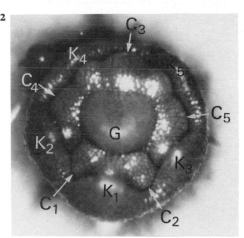

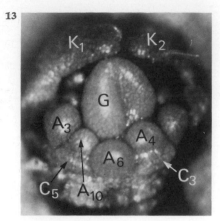

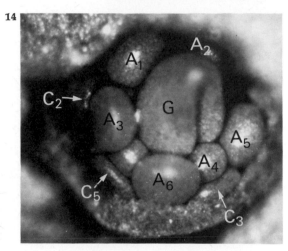

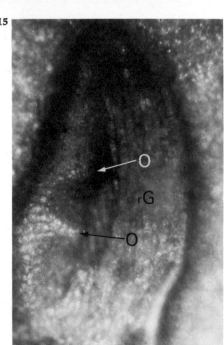

stages of development this slight amount of basal growth is no longer detectable.

Adjacent to each of the first two petal primordia an antesepalous stamen primordium is initiated (A2 and A3 in 6). At about the same time – or perhaps immediately afterward – the primordia of the standard (C3) and the stamen opposed to it (A4) are formed. These two primordia have a common base (7, 10, 11). It is difficult to decide whether this common base is due to inter-primordial growth or whether the petal and stamen primordia are formed on one common hump. If the former assumption is correct, the growth between the petal and stamen primordia would start immediately after or even during the inception of these two primordia. The primordia of the wings (C4 and C5) appear shortly after that of the standard or perhaps even simultaneously with the latter (9). Immediately afterward, the remaining six stamen primordia (A5-10) are initiated (11, 12). It is possible that they appear in very rapid succession two at a time in the following sequence: first the last two antesepalous stamen primordia (A5 and A6), then the two stamen primordia opposite the keel petals (A7 and A8), and, finally, the two stamen primordia opposite the wing petals (A9 and A10). However, this sequence cannot be ascertained. The six stamen primordia might appear simultaneously as is assumed above in 'Sequence of Primordial Inception.' The four stamen primordia opposite the wing and keel primordia are formed on the adaxial side of these primordia. Due to this origin, they appear on a common base with the wing and keel primordia (12, 13). This common base does not grow conspicuously. Therefore, it is no longer visible in later stages of development. The two stamen primordia (A5 and A6) adjacent to the standard primordium (C3) are initiated at the flanks of the C3 A4-primordia (10). As a result of this proximity in origin, and/or due to some slight growth at the base of these

primordia, they form one unit like A1, C1 and C2 in figures 6, 8, 9, 11. With regard to the centers and the outer margins, the antepetalous stamen primordia are inserted inside the antesepalous ones, i.e. the androecium develops in a diplostemonous manner. However, the inner margins of antepetalous and antesepalous stamen primordia are at about the same level (12).

The inception of the gynoecium is a gradual process, and therefore difficult to indicate in the very rapid succession of initiating petals and stamens. One can observe, however, that the gynoecial primordium appears distinctly before the inception of the last six stamen primordia (A5-10), perhaps even before C4 and C5 (7). The gynoecial primordium is horseshoe-shaped and highest on the side toward K1 (10, 11). As it grows upward, the margins come in contact and fuse eventually (13, 14, 21). Usually 3-4 ovule primordia are initiated near each margin in rapid acropetal succession (15, 16). Each ovule primordium forms two integuments (17, 18). During the development of the integuments, the ovules become campylotropous (18).

While the ovules develop, growth begins between 9 of the stamen primordia. As a result of this growth, 9 stamens become inserted on a staminal tube (21), whereas the stamen opposite the standard remains free (20).

Some growth occurs also in a zone underneath the perianth and androecium, which leads to a slight indication of perigny hardly noticeable in the mature flower and therefore not symbolized in the floral formula.

The postgenital fusion of the two keel petals, which gives the impression of a tetramerous corolla in the mature flower, is remarkable (19, 22).

OTHER AUTHORS

I found no literature on the floral development of *Pisum*. However, other genera of the Fabaceae have been studied. Guard (1931) reports on the floral development of the soy bean which has the same floral diagram as the pea. His results differ from my observations with regard to the sequence of primordial inception. He finds acropetal inception of the five whorls of appendages. Within each whorl the order of appearance of primordia is from the abaxial to the adaxial side of the floral apex. Picklum (1954) finds the same sequence of primordial inception in *Trifolium pratense*. Payer (1857), who studied three different palilionaceous genera, also reports acropetal inception of the appendages. Frank (1876), Goebel (1884), and other authors quoted by Frank (1876) note that diverse deviations from acropetal primordial inception occur in different genera of the Fabaceae. For example, Goebel (1884, p. 309) observes that the carpel primordium originates before the inception of all stamen primordia is completed. Eichler (1878, p. 516) reports that the first primordia of any one whorl are initiated before the last primordia of the preceding whorl. Frank (1876) devises a model to accommodate all of these and other deviations within a broader concept of acropetal inception. He

distinguishes two directions of inception: an acropetal and a transversal one. The acropetal direction proceeds from the base of the apex (or bud) to its tip; the transversal direction goes from the abaxial side toward the adaxial side of the apex (or bud). Depending on the relative speed of inception in the two directions, diverse sequential patterns result. For example, in *Vicia cracca* Frank (1876) describes a developmental stage in which the abaxial sepal, stamen and carpel were initiated in acropetal sequence, but only two lateral sepal primordia were present at that time, i.e. the abaxial acropetal direction preceded almost completely inception in the transversal direction. It is highly questionable whether my observations on *Pisum sativum* can be accommodated within Frank's model. To reach a final decision in this matter, the inception of C_3 and A_4 in relation to other primordia should be known more accurately.

Frank (1876) notes also that the primordia of petals and inner stamens originate from common primordia. I think it is more accurate to say that the primordia of the inner stamens appear on the adaxial side of the petal primordia. Since the adaxial side of the petal primordia cannot be sharply delimited from the young receptacle, it may be futile to argue whether inner stamen primordia are formed on the petal primordia or the receptacle. In the case of the adaxial petal (C_3) and the stamen opposite to it, the situation may be different and even more complicated.

Schüepp (1911) and Schumann (1890) describe the floral development of a number of *Lathyrus* and *Vicia* species.

After treatment of young pea plants with phenylboric acid, Haccius and Wilhelmi (1966) obtained floral abnormalities such as open pistils and pistilloid sepals which resemble certain mutants described by Lamprecht and Gottschalk (1964; quoted by Haccius and Wilhelmi 1966).

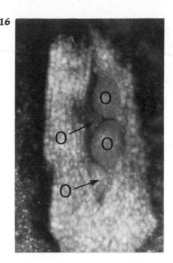

16

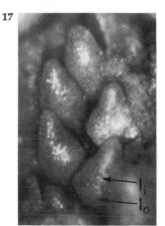

17

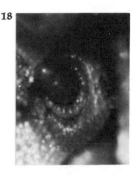

18

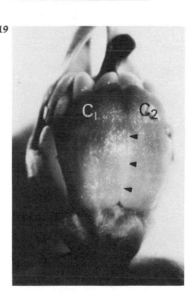

19

Side view of floral bud from which three sepal primordia were removed.
Side view of floral bud from which all sepal primordia were removed.

FIGURES 15-18. Portions of pistils in different stages in development showing ovule (O) inception and formation of integuments (I_o and I_1). x 146
Pistil from which almost half of the wall was removed (rG) to show two ovule primordia (O).
Portion of dissected young pistil showing two ovule primordia near each margin.
Five ovules showing the simultaneous inception of inner (I_1) and outer integument (I_o).
More advanced ovule primordium.
Abaxial view of floral bud from which sepals were removed to show postgenital fusion (arrowheads) of the two keel petals (C_1 and 2). x 10

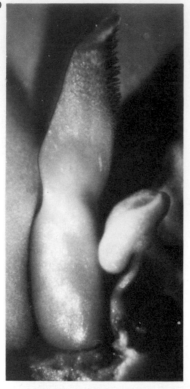
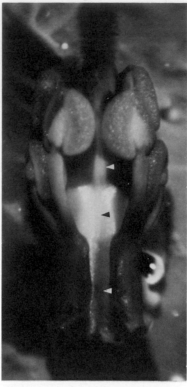

Eichler, A. W. 1878. *Blüthendiagramme*. 2. Teil. Leipzig: Engelmann.

Frank, A. B. 1876. Über die Entwicklung einiger Blüthen, mit besonderer Berücksichtung der Theorie der Interponirung. Jahrb. wiss. Bot. *10*: 204-243.

Guard, A. T. 1931. Development of floral organs of the soy bean. Bot. Gaz. *91*: 97-102.

Goebel, K. 1884. *Vergleichende Entwicklungsgeschichte der Pflanzenorgane*. Berlin: R. Friedländer und Sohn.

Haccius, Barbara und Wilhelmi, D. 1966. Mutationen kopierende Blüten-Anomalien bei *Pisum sativum* nach Phenylborsäure-Behandlung. Planta *69*: 288-291.

Payer, J.B. 1857. *Traité d'organogénie comparée de la fleur. Texte et Atlas*. Paris: Librairie de Victor Masson.

Picklum, W. E. 1954. Developmental morphology of the inflorescence and flower of *Trifolium pratense* L. Iowa State Coll. J. Sci. *28*: 477-495.

Schüepp, O. 1911. *Beiträge zur Entwicklungsgeschichte der Schmetterlingsblüte*. Thesis, Eidgen. Polytechn. Schule, Zürich, Switzerland.

Schumann, K. 1890. *Neue Untersuchungen über den Blüthenanschluss*. Leipzig: W. Engelmann.

20 Side view of pistil and adaxial stamen (A4) in a more advanced bud. x ca. 37

21 Adaxial view of nearly mature flower after removal of sepals, petals, and adaxial (free) stamen. Arrowheads point to the postgenital fusion line in the pistil. x ca. 28

22 Mature flower. x ca. 4

20 / GERANIALES (GRUINALES)

GERANIACEAE

Pelargonium zonale (L.) L'Hérit. ex Aiton
(zonal or horseshoe geranium)

Floral diagram
Note: Variable aestivation of the petals occurs in most mature flowers.

Floral formula: $\cdot | \cdot$ K5 C5 A(5 + 5) $\underline{G}$(5) O10
Note: Three of the antepetalous stamens are staminodial.

Sequence of primordial inception
K1,2,3,4,5, C1-5, A_k1-5, A_c1-5, G1-5, O1-10
Note: The primordia of the corolla and the two androecial whorls probably arise in a very rapid succession which approaches simultaneity.

DESCRIPTION OF FLORAL ORGANOGENESIS

The sepal primordia are initiated successively in a quite unusual pattern (1, 2): the first and the second sepal primordia have the same position as in the usual spiral pattern; but the third sepal primordium is formed adjacent to the second one continuing the spiral direction of the first and second primordia. The fourth and fifth sepal primordia are initiated in such positions that they form a spiral whose direction is opposite to that of the first and second primordia. Therefore, when the first two

sepal primordia are formed in counterclockwise direction, the last two appear in clockwise direction (1) and vice versa (2, 3). Often it is difficult to distinguish between the second and third sepal primordia. Hence, other patterns may occur besides the one described above. It is not known whether the normal spiral sequence of inception occurs in at least some buds.

The petal primordia are initiated at about the same time. It is possible that they actually appear in a very rapid sequence which approaches simultaneity (2). In contrast to the sepal primordia and the primordia of the androecium and gynoecium, they grow quite slowly in the earlier stages of development (4-9).

After petal inception, the antesepalous stamen primordia are initiated (3). Whereas the first three of them are of about the same size, the fourth and the fifth are slightly smaller. This slight size difference probably indicates a very rapid sequential origin that approaches simultaneity. After the antesepalous stamens, the antepetalous stamens are initiated centrifugally, i.e., at a level between that of the petal primordia and primordia of the antesepalous stamens (4-5). Consequently, the androecium originates in

---

1 Top view of floral apex showing inception of sepals (K). The numbers indicate the sequence of inception. x 146

2 Top view of floral bud showing sepal (K) and petal (C) primordia. x 146

3 Top view of floral bud with the primordia of the perianth and androecium. Only the sepal primordia are labelled according to their sequence of inception. The antesepalous stamen primordia are well formed. The petal primordia are hardly visible, and the antepetalous stamen primordia are not yet very distinct. x 146

4 Side view of a floral bud from which the sepal primordia were removed to show the primordia of the antesepalous stamens (A_K), two petals (C), and antepetalous stamens one of which is labelled (A_c). x 146

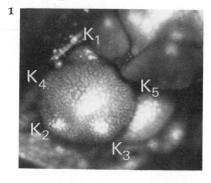

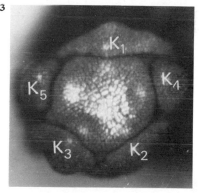

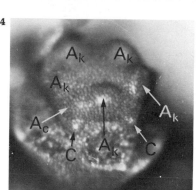

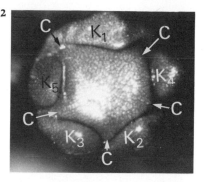

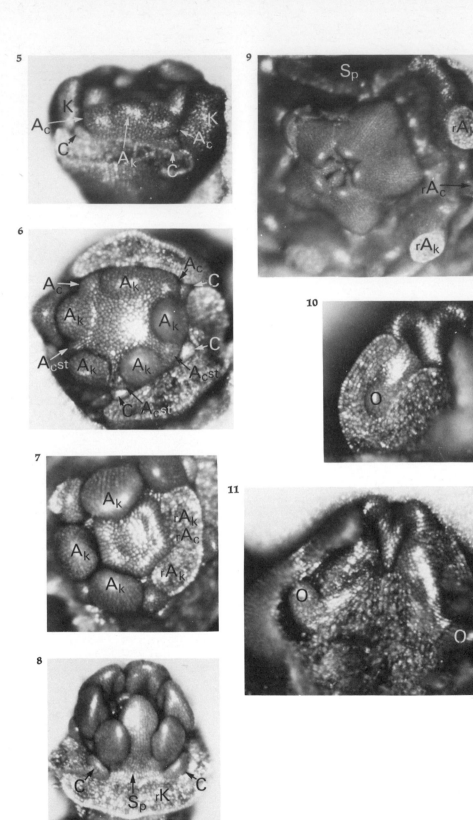

an obdiplostemonous manner. Three of the antepetalous primordia grow more slowly than the other two which are adjacent to both margins of the first sepal primordium (6, 7). The more slowly growing primordia develop into filament-like staminodia, whereas the other two, like all the antesepalous primordia, develop into stamens with a filament and an introrse anther. There is considerable upgrowth in a ring-zone underneath the primordia of the seven stamens and the three staminodia. As a result, a high androecial tube is formed (9).

After androecial inception, usually five gynoecial primordia are formed opposite the antepetalous (outer) stamen primordia. Growth extends immediately into the region between the five gynoecial primordia. As a result, a five-lobed pentagonal ridge is formed (7). The gynoecial primordia on the ridge eventually develop into the styles with stigmas. The cylindrical portion of the ridge develops into the ovary. Five septa grow out at positions alternating with the gynoecial primordia in continuity with the floral apex. Thus, five locules are formed. In each locule two ovules are initiated on the young septa near the center of the ovary (10, 11). Each ovule becomes bitegmic and anatropous.

After gynoecial inception, an area between the first sepal primordium and the stamen primordium superposed to it grows much less than the other portion of the developing flower. As a result, a depression, i.e. the spur, is formed which becomes very deep in the receptacle and pedicel of the mature flower (13).

OTHER AUTHORS

Sauer (1933) describes the external morphology and anatomy of the mature flower and very late developmental stages in this species. He interprets the spur as a formation of the sepal. This interpretation is due to a different delimitation of sepal and receptacle. There is no discrepancy in the observations of Sauer and myself. The same is true for the developmental study of the spur by Labbe (1964).

Payer's (1857) report on *Pelargonium inquinans* agrees with my observations with two exceptions: he observes normal spiral (quincuncial) sepal inception, and that the spur is formed opposite the second sepal.

Recently, Eckert (1966) has described the early floral development of *Geranium nodosum* L. especially with regard to the problems of obdiplostemony. She finds that the antesepalous and antepetalous stamen primordia are formed at about the same level. The obdiplostemony of the mature flower is due to secondary positional changes during later developmental stages. With regard to sepal inception, she reports a usual spiral 2/5 sequence. Apart from these differences, *Geranium nodosum* and *Pelargonium zonale* show very similar floral development in the early stages.

Hieke (1963) presents a survey of all abnormalities in this species that have been observed by himself and other authors.

BIBLIOGRAPHY

Eckert, Gertrude. 1966. Entwicklung-schichtliche und blütenanatomische Untersuchungen zum Problem der Obdiplostemonie. Bot. Jahrb. *85*: 523-604.
Hieke, K. 1963. Morfologické abnormality in *Pelargonium zonale* Ait. Acta pruhon. *5*: 1-12.
Labbe, A. 1964. Sur l'éperon de la fleur de *Pelargonium*. Bull. Soc. Bot. France (Paris) *111*: 321-324.
Payer, J.B. 1857. *Traité d'organogénie comparée de la fleur. Texte et Atlas.* Paris: Librairie de Victor Masson.
Sauer, H. 1933. Blüte und Frucht der Oxalidac., Linac., Geraniac., Tropaeolac., und Balsaminac. Planta *19*: 417-481.
Zanker, J. 1930. Untersuchungen über die Geraniaceae. Planta *9*: 681-717.

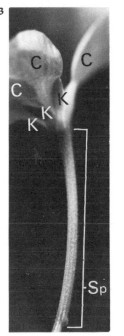

5 Side view of a slightly older bud than that of figure 4 showing clearly the obdiplostemonous origin of the androecium. The sepal primordium opposite the labelled antesepalous stamen primordium (A_K) was removed. x 146

6 Top view of a floral bud from which four of the sepal primordia were removed to show three of the five petal primordia (C), the antesepalous stamen primordia (A_K) and the antepetalous stamen primordia (A_c) three of which will form only staminodia (A_cst). x 146

7 Top view of inner portion of a floral bud showing inception of five gynoecial primordia and growth between them thus forming a five-lobed pentagonal ridge. Two of the antesepalous stamen primordia (rA_K) and one of the staminodial antepetalous stamen primordia (rA_c) were removed. x 146

8 Side view of a floral bud from which the young calyx was removed to show the area of spur formation (Sp). x 85

9 Top view of a floral bud from which the young sepals and stamens were removed to show the young gynoecium and a portion of the androecial tube to the right. Sp is the area of spur formation. x 85

1 Side views of dissected gynoecia showing stages of ovule (O) inception and development. Two gynoecial primordia are visible in each figure. x 146

3 Mature flowers. Figure 12, top view. Figure 13, side view. The bracket indicates the length of the spur (Sp). x ca. 2

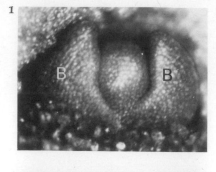

1

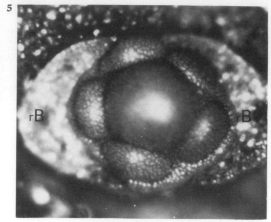

5

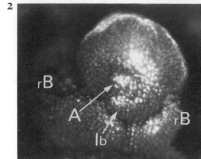

2

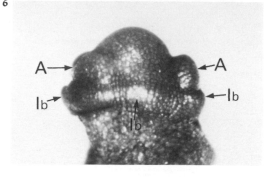

6

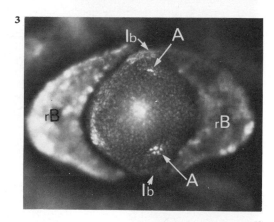

3

7

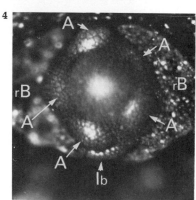

4

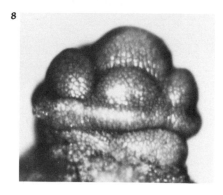

8

21 / GERANIALES

EUPHORBIACEAE

Euphorbia splendens Boj. ex Hook.
(= **E. millii** Desmoul.)
(crown of thorns)

Cyathial diagram

Cyathial formula: ✳ B(2) (Nectary Glands 5
Involucral Bracts 5) Scales 10 A-Groups 5:A(5) per
group D(3) G̲(3) O3

Sequence of primordial inception
See text.

DESCRIPTION OF CYATHIAL ORGANOGENESIS

The two large red bracts are initiated first
and almost cover the cyathial apex before
further development becomes obvious (1).
Then the primordia of the involucral bracts
are formed in a rapid spiral sequence (2).
Immediately after their inception, interprimordial growth occurs, thus leading to
the formation of the involucral tube (5-9).
Later on the five nectary gland primordia
are formed on the upper rim of the involucral tube in alternation with the involucral
bracts (12, 13, 28). Whereas the nectary
glands grow upward as simple fleshy structures, the involucral bracts grow inward
and become lobed (28).

The first five androecial primordia which
are formed are opposite the primordia of
the involucral bracts (2-9). Each of them
appears immediately after the subtending

involucral bract primordium (or in some cases even simultaneously with it). Consequently the first androecial primordium may be initiated before the fifth involucral bract primordium (3). In the same sequence as the first five androecial primordia, a second set of five androecial primordia is initiated. Each primordium of the second set is formed centrifugally on the side of a primordium of the first androecial set (10-13). The primordia of the second set are all formed on the right side or all on the left side of the first primordia (13). However, in many buds one of the primordia of the second set is formed on the opposite side than the other four; thus, two of the primordia of the second set are side by side (11, 12). A third set of androecial primordia is formed centrifugally on the outer side of the primordia of the first set. In the

---

1 Side view of cyathial apex with the two primordia of the red bracts (B). x 146

FIGURES 2-13. Cyathial buds from which the primordia of the two red bracts were removed (rB). x 146

-3 Slightly older stages than that of figure 1. Inception of involucral bracts (Ib) and the first stamens (A). Figure 2, side view. Figure 3, top view.

-5 Top views of cyathial buds showing inception of the first five stamen primordia. In figure 5 inception of the second set of stamens and scales might be faintly visible.

-9 Side views of cyathial buds showing inception and development of involucral bract primordia (Ib) and tube. The first set of stamen primordia (A) is visible; the second set is not yet clearly discernible.

-3 Top views of cyathial buds showing inception of the second set of stamens (A) and the scales (S). In figure 10 the arrow (A) points to the first stamen primordium of the second set. In figures 12 and 13 the young gynoecium was removed (rG). Ib = primordia of involucral bracts; Gl = primordia of glands which alternate with those of the involucral bracts on the involucral tube.

4 Abaxial view of two androecial groups and scales (S). The stamen primordia in each group are labelled according to the order of inception. Involucral tube was removed (rIt). x 146

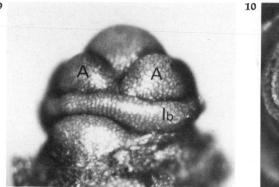

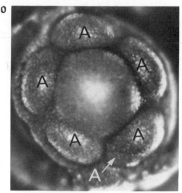

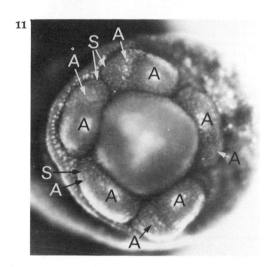

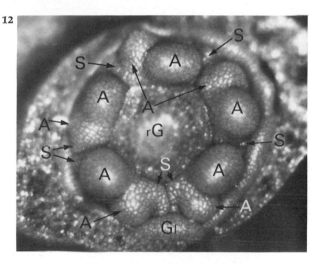

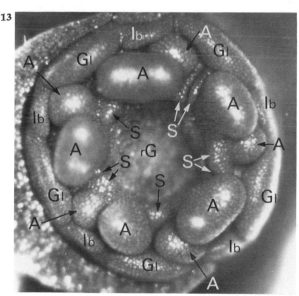

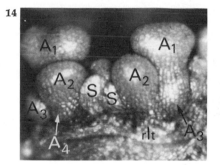

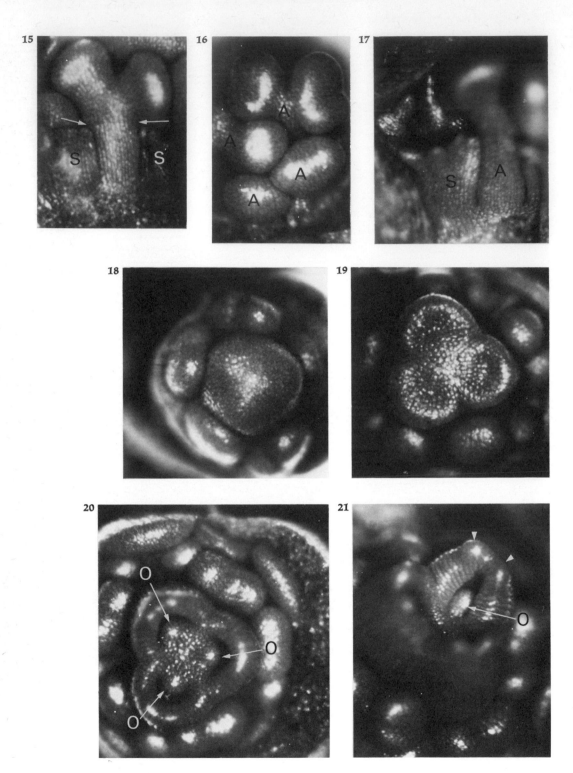

same manner a fourth set of stamen primordia is initiated on the stamen primordia of the second set (14) and finally a fifth set of stamen primordia appears on those of the third set. Thus five groups of five androecial (stamen) primordia are formed. In each group the androecial primordia are arranged in a zig-zag pattern (16). Each of the androecial primordia develops first two lobes (13, 15), and then each lobe becomes again two-lobed (16). As a result, a stamen with an extrorse anther is formed whose thecae are widely separated from each other (16, 30). On the upper part of the filament growth becomes restricted in a ring-shaped zone (15).

In close association with the androecial primordia are scales which develop on either side of each androecial group. These scales are initiated at approximately the same time as the second set of androecial primordia (10, 11). They grow from the base on both sides of the first five androecial primordia around the androecial groups, fusing postgenitally with the involucral tube. In later stages the scales branch profusely, and at maturity become entangled in the stamens about which they grow (17). Little growth occurs underneath the stamens of each group and the associated scales and involucrae bract. This is not indicated in the floral formula and diagram.

The gynoecium is initiated at about the same time as the second set of androecial primordia (18). The cyathial apex accelerates its growth in three areas, thus forming a triangle of three horseshoe-shaped gynoecial primordia. They grow up in spatial continuity with each other and three septa, thus forming a three-locular ovary (19-22). Each gynoecial primordium branches dichotomously at its tip. As a result six styles and stigmas are formed on the trimerous gynoecium (20, 21, 28, 29). At about the time when the first five androecial primordia begin to branch at their apices, the three ovules are initiated simultaneously on the flanks of the central apex (19, 20, 21). The floral apex (i.e. the placenta) grows upward, carrying the young ovules up along with it (21, 22). The two integuments of the ovules are initiated in very close succession (23, 24). Above each ovule a primordium develops which is said to function as an obturator (24, 25).

When the gynoecial primordia branch at their apex, a crescent-shaped primordium is initiated below each gynoecial primordium (26, 27). Growth extends eventually between the margins of these three primordia which have been classified as disc (D) or perianth depending on the interpretation of the cyathium as a flower or inflorescence respectively.

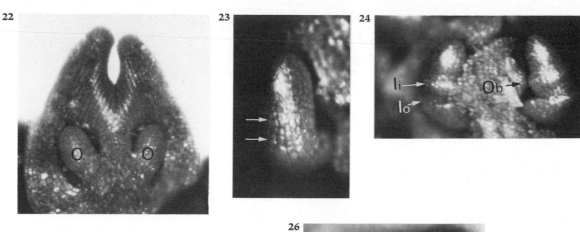
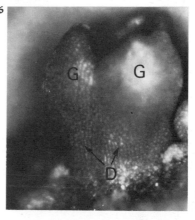
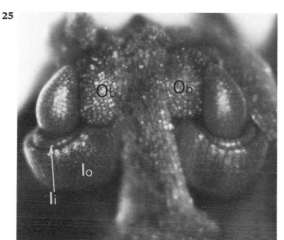
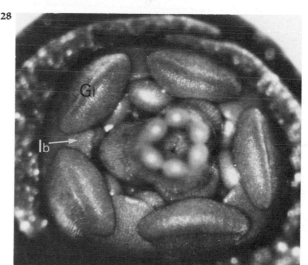
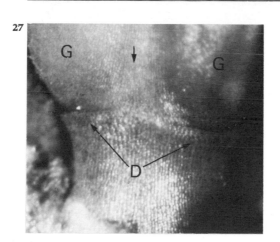

Stamen primordium showing origin of constriction (see arrows). Adaxial view. x 146
Abaxial view of an androecial group of four stamen primordia (A) in different developmental stages. x 146
Branching of a scale primordium (S). Adaxial view. x 146
Top views showing inception of three gynoecial primordia and three ovules (O in figure 20). x 146
Side view of more advanced gynoecium showing one locule with an ovule primordium and dichotomous branching of the apex of the gynoecial primordia (see arrowheads). x 146
Side view of longitudinally dissected young gynoecium showing insertion of two ovule primordia (O). x 146
Side view of ovule primordium. The two arrows indicate the sites of integument inception. x 246
Side views of a portion of the central placenta with two ovule primordia showing stages of integument development (I_o = outer integument, I_i = inner integument). Ob = primordium of the obturator. x 146
Side view of a young gynoecium in the developmental stage of that in figure 21 showing inception of disc (tepal) primordia (D). x 146
Side view showing an older stage of disc (perianth) development. Arrow indicates the interior position of the septum between two gynoecial primordia. x 85
Top view of a young cyathium after removal of the two bract primordia. One of the involucral bracts already shows lobing (see arrow). One of the five glands is labelled (Gl). x 85

29

30

Payer (1857) describes the organogenesis of the cyathium in *Euphorbia lathyris*. His findings correspond closely to ours with few exceptions. He reports five scales only which he describes as disc. They originated after ovule inception and developed between the androecial primordia without any special connection to the latter.

Schmidt (1907) who quotes and discusses other older literature (except Lyon's (1898) developmental study of *Euphorbia corollata*) describes the floral development of *Euphorbia splendens*. His illustrations which are based on serial sections are matched by our observations; but he mentions some features which we could not confirm. He claims that the primordia of the first five stamens are formed before involucral inception. And he mentions that the involucral bracts do not follow exactly the spiral sequence in which the first five stamen primordia are formed.

Michaelis (1924) who also studied *Euphorbia splendens* reported observations which contradict our findings. He claims that the third stamen primordium in each group is formed on the second stamen primordium, whereas it was observed by Schmidt and myself that the third stamen primordium arises on the first one (see figure 14). Furthermore, Michaelis mentions – for *Euphorbia salicifolia* – that each involucral bract and the superposed first stamen originate from one common primordium (figures 8-10 on his plate 40).

Haber (1925), though not discussing cyathial development directly in her article on the anatomy and morphology of the cyathia of various species of *Euphorbia*, does recognize that the development of the scale is in some species closely associated with the androecial primordia. For example *Euphorbia portulacoides*, *Euphorbia darlintonii*, and *Euphorbia buxifolia* are described and illustrated by her as cases in point. Bodmann's (1937) comparative study is on mature structures.

Struckmeyer and Beck (1960) and Goddard (1961) discuss flower initiation and development in *Euphorbia pulcherrima* (Poinsettia), but bring out little new on its cyathial development. What little they do on the development of the cyathium does not contradict any of our observations on *Euphorbia splendens*.

BIBLIOGRAPHY

Bodmann, H. 1937. Zur Morphologie der Blütenstände von *Euphorbia*. Öster. bot. Zschr. *86*: 241-279.

Goddard, G.B. 1961. Flower initiation and development in the poinsettia (*E. pulcherrima*). Proc. Amer. Hort. Sci. *77*: 564-571.

Haber, J.N. 1925. The anatomy and morphology of the flower of *Euphorbia*. Ann. Bot. *39*: 657-707.

Lyon, F.M. 1898. A contribution to the life history of *Euphorbia corollata*. Bot. Gaz. *22*: 418-427.

Michaelis, P. 1924. Blütenmorphologische Untersuchungen an den Euphorbiaceen ... Goebel's Bot. Abhandl. *3*: 1-150.

Payer, J.B. 1857. *Traité d'organogénie comparée de la fleur. Texte et Atlas*. Paris: Librairie de Victor Masson.

Schmidt, H. 1907. Über die Entwicklung der Blüten und Blütenstände von *Euphorbia* L. und *Diplocyathium* n.g. Beih. Bot. Centralbl. *22*: 21-68.

Struckmeyer, B.E. and Beck, G.E. 1960. Flower bud initiation and development in poinsettia. Amer. Soc. Hort. Sci. *75*: 730-738.

29 Top view of mature cyathium with receptive stigmas. x ca. 6
30 Top view of mature cyathium at anthesis. x ca. 6

22/SAPINDALES

ANACARDIACEAE

Rhus typhina L. (staghorn sumac)

Floral diagram

Floral formula: ✳ [K5 C5 A5 D(5)]G̲(3) O1

Sequence of primordial inception
K1,2,3,4,5, C1-5, A1-5, G1,2-3, O1, D
Note: The primordia of the corolla and the
androecium may arise in very rapid succession
which approaches simultaneity.

DESCRIPTION OF FLORAL ORGANOGENESIS

The five sepal primordia are formed in a
distinct spiral sequence (1). Soon after the
inception of the fifth sepal primordium, the
five petal primordia are initiated at about
the same time (1-3). Then the five stamen
primordia appear (3-4). Each of the stamen
primordia forms a stamen with an introrse
anther.

---

1 Top view of floral bud showing spiral inception of
  sepal primordia (K) and the first indication of
  petal inception (C). x 146
2 Side view of a floral bud showing sepal and petal
  primordia. x 146
4 Top views of floral buds showing primordia of
  sepals (K), petals (C), and stamens (A). In figure
  4 three of the stamen primordia are hidden below
  sepal primordia. x 146

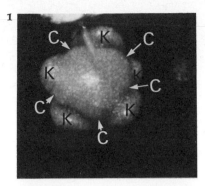

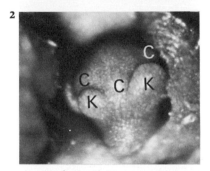

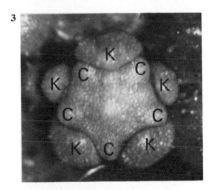

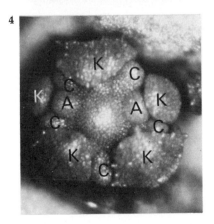

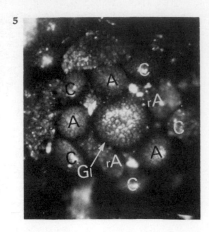

5

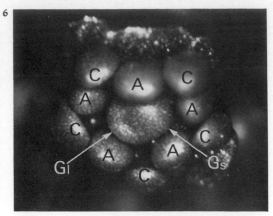

6

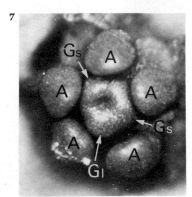

7

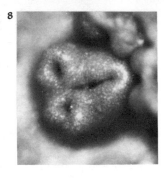

8

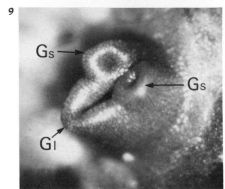

9

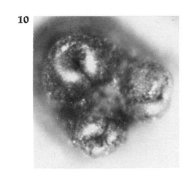

10

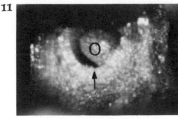

11

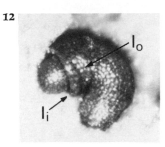

12

After stamen inception, the floral apex enlarges considerably (4). Then one large crescent-shaped gynoecial primordium arises (5-6), followed by two small gynoecial primordia (6-7). Growth between the margins of the three gynoecial primordia leads to the formation of a cylindrical portion which will constitute the ovary. In contrast to the large gynoecial primordium which remains crescent-shaped, each of the small ones becomes tubular due to restriction of growth in its center (8-9). As a result, there are two rudimentary chambers in addition to the large ovarial cavity. Each of the three gynoecial primordia forms a style (10). There is hardly any difference in the size of styles, i.e., the styles formed by small gynoecial primordia are approximately as large and of the same structure as the style formed by the large gynoecial primordium (13, 14). Only one ovule primordium is formed near the base of the ovary between the two small gynoecial primordia (11). It becomes bitegmic and anatropous (12).

A conspicuous fleshy disc arises in the later stages of development. It becomes five-lobed and assumes an orange colour in the mature flower (14).

Some growth occurs in a zone below the developing perianth, androecium, and disc. As a result of this, the flower becomes slightly perigynous. This perigyny is only indicated in the floral formula, not in the diagram.

OTHER AUTHORS

I found no recent literature on the floral development of *Rhus*, except an abstract by Hall (1947) that is in agreement with my description. Payer (1857) describes the floral organogenesis of *Rhus coriaria*. His results agree rather well with my observations on *Rhus typhina*. Grimm (1912) studied the floral development of the dioecious *Rhus toxicodendron*. He reproduces cross-sections through developing gynoecia which illustrate clearly the pseudomonomerous condition in the ovarial region. McNair (1921) gives a very sketchy and brief account of the floral development in *Rhus diversiloba* which does not contradict my findings. Eckardt (1937) demonstrates the two vestigial locules in the upper portion of the ovary of *Rhus typhina*. Hence, my observations concur with the conclusions of other investigators.

BIBLIOGRAPHY

Eckardt, Th. 1937. Untersuchungen über Morphologie, Entwicklungsgeschichte und systematische Bedeutung des pseudomonomeren Gynoeceums. Nova Acta Leopold. N.F. *5*: 1-112.

Grimm, J. 1912. Entwicklungsgeschichtliche Untersuchungen an *Rhus* und *Coriaria*. Flora *104*: 309-334.

Hall, B.A. 1947. The nature of the ovary in the genus *Rhus*. Amer. J. Bot. *54*: 584 (Abstract).

McNair, J.B. 1921. The morphology and anatomy of *Rhus diversiloba*. Amer. J. Bot. *8*: 179-192.

Payer, J.B. 1857. *Traité d'organogénie comparée de la fleur. Texte et Atlas.* Paris: Librairie de Victor Masson.

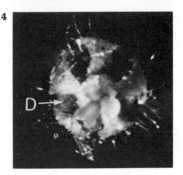

Top view of a floral bud showing inception of the large gynoecial primordium (Gl). Primordia of sepals and two stamens (rA) were removed. x 146

Floral bud from which sepal primordia were removed to show inception of gynoecium. The large gynoecial primordium (Gl) and one of the small ones (Gs) are visible. x 146

Top view of floral bud from which sepal and petal primordia were removed to show the three gynoecial primordia and the stamen primordia. x 146

Two different views of young gynoecia showing the development of the ovary and the large and small gynoecial primordia. x 146

Top view of young pistil showing development of the three styles. x 146

Dissected young ovary showing the single ovule primordium (O). Arrow indicates base of the ovary. x 146

Young ovule with primordial inner integument (I_i) showing inception of outer integument (I_o). x 146

Mature flowers. In figure 14 the petals and stamens were removed to show the disc (D). In figure 13 the corolla exhibits a slight degree of zygomorphy which is not characteristic of all flowers in this species. Figure 13, x ca. 10. Figure 14, x ca. 18.

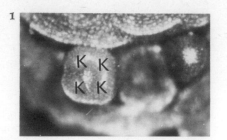

1

2

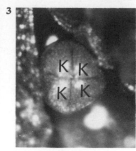

3

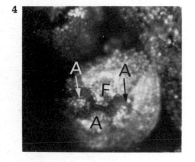

4

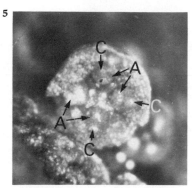

5

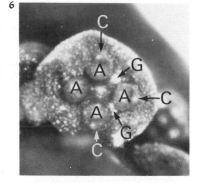

6

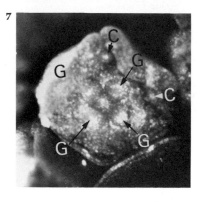

7

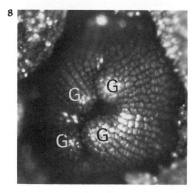

8

23 / RHAMNALES

RHAMNACEAE

Rhamnus cathartica L.
(common buckthorn)

Floral diagram

Floral formula: $*$ [K(4) C4 A4] $\underline{G}$(4) O4
Note: This species is dioecious.
In the male flower the pistil is rudimentary, whereas
in the female flower the stamens and petals are
rudimentary. The inner side of the calyx tube (cup)
functions as disc.

Sequence of primordial inception
K1-4, A1-4, C1-4, G1-4, O1-4

DESCRIPTION OF FLORAL ORGANOGENESIS

Differences in the organogenesis between
the male and female flowers occur only in
late stages of development, after the inception and early development of the pistil.
Therefore, the floral organogenesis will
first be described as it applies to flowers of
both sexes, and then the divergence in the
late development will be pointed out.

On the floral apex, the sepal primordia
are initiated more or less simultaneously
(1, 2, 13). Soon after their inception, growth
starts in the regions between them, thus
leading to the formation of a calyx tube
(1-3, 13, 14).

When the sepal primordia begin to cover
the floral apex (3, 14), the stamen primordia
are initiated at about the same time on the
developing calyx tube (4, 15). At this stage

of development, the floral apex is concave; hence, it can be delimited from the young calyx tube only arbitrarily.

The petal primordia are initiated soon after the inception of the stamen primordia on the abaxial (outer) base of the latter (5, 16). They grow much more slowly than the stamen primordia (6, 17, 18, 23), but immediately after their inception they are about half as big as the stamen primordia (6, 15). Thus, each stamen primordium with the opposed petal primordium gives the impression of one two-lobed primordium with unequal lobes (5, 16).

Immediately after the inception of the petal primordia, the four gynoecial primordia appear at the flanks of the concave floral apex, alternating with the four stamen primordia (6, 7, 16). Growth extends

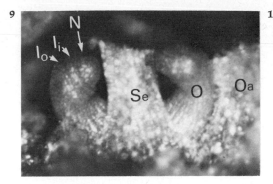

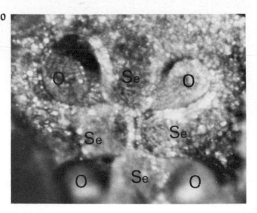

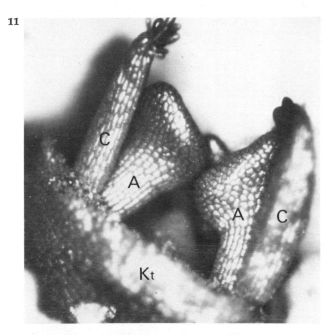

FIGURES 1-12. Developmental stages of female flowers.

3 Top views of floral buds showing inception and early development of sepals (K) and calyx tube. x 146

4 Floral bud from which the young calyx and one stamen primordium were removed to show floral apex (F) and the other three stamen primordia (A). x 146

6 Floral buds from which the young calyx was removed to show inception of petals (C) and gynoecial primordia (G). x 146

7 Floral bud from which the young calyx, all stamen primordia and two petal primordia were removed to show the gynoecial primordia (G). x 146

8 Top view of young pistil showing the four gynoecial primordia (G) on top of the young ovary wall. x 146

9 Lower portion of longitudinally dissected ovary showing two ovule primordia (O) with incipient integuments (I_o, I_i) at the base of septum (Se) and ovary wall (Oa). x 146

10 Portion of a transversely dissected ovary showing postgenital fusion of the four septa (Se) and parts of ovules (O). x 146

11 Side view of the upper portion of the calyx tube (Kt) from which the sepals were removed to show nearly mature petals (C) and staminodia (A). x 146

12 Top view of mature female flower. x ca. 12

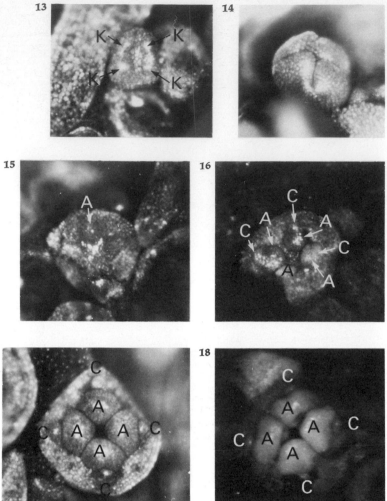

to the regions between the gynoecial primordia; in this way the formation of the pistil wall is initiated. During the upgrowth of the developing pistil wall, four septa are formed in alternation with the gynoecial primordia. The septa which grow up from the floor of the young ovary eventually fuse postgenitally in the center of the ovary (10, 20-22). Thus, the ovary is subdivided into four locules. At the base of each locule one ovule is initiated. Each ovule becomes bitegmic and anatropous (9). The gynoecial primordia grow up and fuse postgenitally in their lower portion, forming one style and four terminal stigmas (12).

In the female flower no fertile stamens are formed; instead, small staminodia develop (11, 12). The petals also remain very small. They are nearly radial in symmetry due to an almost complete lack of marginal growth in the petal primordia (11).

In the male flower the pistil remains very small, and does not function, although ovules are formed as in the female flower (22). Floral buds with only three gynoecial primordia were found (20). The petals remain smaller than the stamens, but they become quite broad and surround the stamens due to much marginal growth (23, 24).

OTHER AUTHORS

Bennek (1958) describes the development of perianth and androecium of *Rhamnus cathartica* and other species and genera of the Rhamnaceae. She reports (p. 431) that the petal primordia appear on the outer (abaxial) side of the stamen primordia. This agrees with my observations, especially if one realizes that the outer side of the stamen primordium cannot be clearly delimited from the young calyx tube. However, in the legend to figure 4 of plate 23, Bennek refers to the origin of stamens and petals as a process of splitting of one primordium. I do not agree with this description, since one primordium which forms the stamen is initiated before the other primordium which develops into the petal. I realize, however, that in a stage of development like that of my figures 6 and 16 one might get the wrong impression of splitting of one primordium.

Older literature is quoted and discussed by Bennek (1958).

BIBLIOGRAPHY

Bennek, C. 1958. Die morphologische Beurteilung der Staub- und Blumenblätter der Rhamnaceen. Bot. Jb. 77: 423-457.

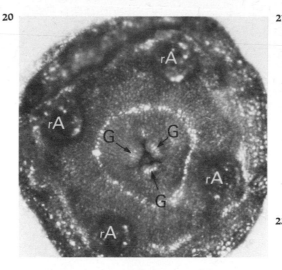

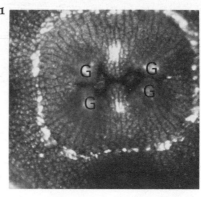

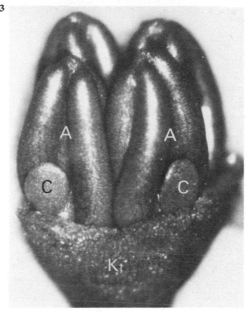

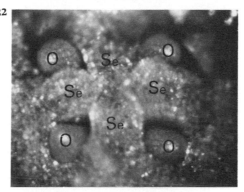

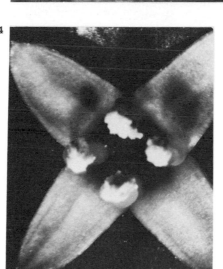

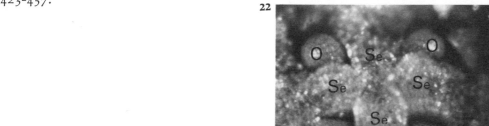

FIGURES 13-24. Developmental stages of male flowers.

14 Floral buds showing inception and early development of sepals (K) and calyx tube. x 146

18 Top views of floral buds from which the young calyx was removed to show inception and early development of stamens (A) and petals (C). In figure 15 petal primordia are not yet discernible. x 146

19 Longitudinally dissected pistil showing two gynoecial primordia. x 146

21 Top views of young pistils. The tips of the septa are visible between the gynoecial primordia (G). Figure 20 shows an unusual bud with only three gynoecial primordia (G); stamen primordia were removed (rA). Figure 20, x 85. Figure 21, x 146

2 Portion of a transversely dissected ovary showing the four septa (Se) and ovule primordia (O). x 146

3 Side view of upper portion of calyx tube (Kt) from which the calyx lobes were removed to show young stamens (A) and petals (C). x 85

4 Top view of mature male flower. x ca. 12

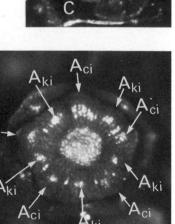

24 / MALVALES (COLUMNIFERAE)

MALVACEAE

Malva neglecta Wallr.
(running mallow)

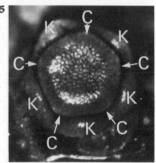

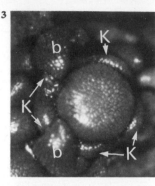

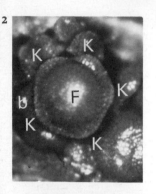

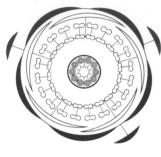

Floral diagram
Note: The contortion of the corolla may be clockwise (as shown) or counterclockwise.

Floral formula:
$$\text{✳ K(5)[C5A(ca. 40-48)] } \underline{G}\text{(ca. 13-15) O ca. 13-15}$$

Sequence of primordial inception
K1-5, C1-5, A_{ci}1-5, A_{ki}1-5, A_{co}1-5, A_{ko}1-5, doubling of previous A's, $G_{primary}$1-5, $G_{secondary}$1-ca. 15, O1-ca. 15
Note:
A_{ci} = inner antepetalous androecial primordium,
A_{ki} = inner antesepalous androecial primordium,
A_{co} = outer antepetalous androecial primordium,
A_{ko} = outer antesepalous androecial primordium,
$G_{primary}$ = first formed gynoecial primordium,
$G_{secondary}$ = gynoecial primordium formed on top of $G_{primary}$

DESCRIPTION OF FLORAL ORGANOGENESIS

After the inception of three bracteole primordia (1), the five sepal primordia are initiated in a very rapid spiral sequence which approaches simultaneity (2). Immediately after sepal inception, the primordial calyx tube appears due to interprimordial growth (1, 2, 8).

The five petal primordia and the first

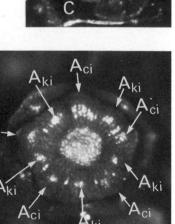

five antepetalous primordia of the androecium seem to appear simultaneously (3-5). However, the possibilities that either the corolla precedes the androecium or vice versa cannot be excluded. Whereas the corolla is initiated as five distinct primordia (5, 11, 12), the androecium appears as a ridge (3, 4). At first this ridge shows five slight elevations opposite the petal primordia (5), and then five additional elevations opposite the sepal primordia (6, 7). Occasionally one androecial primordium of the first series and one of the second series seem to form a pair opposite a petal primordium (7). As the ridge increases in height, the 10 androecial primordia inserted on it grow upward. Meanwhile, outside each of the androecial primordia, another androecial primordium is initiated on the abaxial side of the androecial ridge (9, 10). Probably the five antepetalous primordia are formed first followed immediately by the five antesepalous primordia. Then each of the 20 androecial primordia forms two primordia in a more or less lateral direction (10-13). As a result of this doubling, the final number of 40 androecial primordia is produced (14). Each one will develop into a stamen with two pollen sacs (15, 16). Concomitantly, the androecial ridge, on which they are inserted, grows upward and forms a conspicuous staminal tube (16, 17). In most flowers the development of the androecium is not as regular as described above. Some more androecial primordia may be formed in various ways. Usually these complications arise after the formation of the first ten or twenty androecial primordia, making later developmental stages hard to understand (11, 12). In ten randomly selected flowers the average number of stamens was 43. Only three of the ten flowers had exactly 40 stamens.

Although the petal primordia are initiated on the flanks of the floral apex, they are carried upward on the developing androecial tube due to some growth in a zone below the insertion of the incipient petal primordia (11, 12). In the mature flower, therefore, the petals are attached to the abaxial base of the androecial tube. Each petal becomes two-lobed at its tip (24).

After (or perhaps during) the lateral division of the first 20 androecial primordia, five primary gynoecial primordia are initiated on the same radii as the petal primordia (18). Due to interprimordial growth, the five gynoecial primordia merge into an almost even ridge which surrounds the floral apex (19). Then additional secondary gynoecial primordia appear like

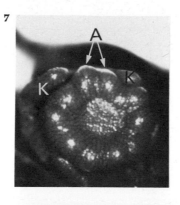

7

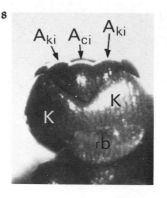

8

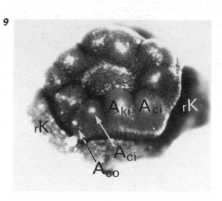

9

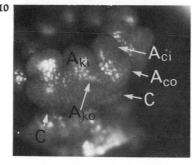

10

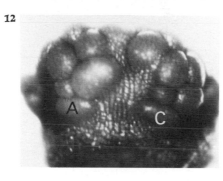

11

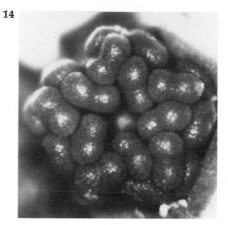

12

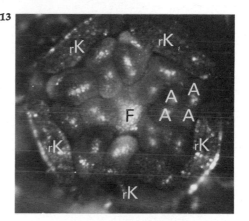

13

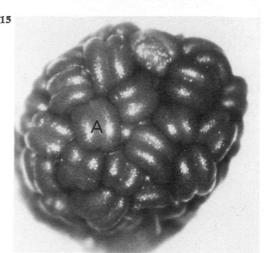

14 15

FIGURES 1-8. Side and top views of young undissected floral buds several of which show the surrounding bracteole primordia (b). x 146

1-2 Sepal (K) inception in side view (1) and top view (2).

3-5 Inception of petals (C) and inner antepetalous androecial primordia in top view (3, 5) and side view (4).

6-8 Inception of inner antesepalous androecial primordia (A_{ki}) immediately after the inner antepetalous ones (A_{ci}). In figure 7 at least two androecial primordia appear as an alternisepalous pair of primordia (see arrows). Figures 6-7 are top views; figure 8 is a side view.

FIGURES 9-15. Floral buds from which the young calyx was removed to show androecial development. x 146

9 Inception of outer antepetalous androecial primordia one of which is labelled (A_{co}).

10 Inception of outer antesepalous androecial primordia one of which can be seen in this side view (A_{ko}).

12 Side views of floral buds after inception of inner and outer androecial primordia which divide(d) laterally. Only the petal primordia (C) are labelled.

13 Top view showing lateral division of the first 20 androecial primordia. Four androecial primordia are labelled (A) which were formed from one inner and one outer antepetalous androecial primordium (A_{ci} and A_{co} in figure 10).

14 Slightly older stage than that of figure 13.

15 Each of the ca. 40 androecial primordia in the stage of figure 14 has formed one bisporangiate stamen, one of which is labelled (A).

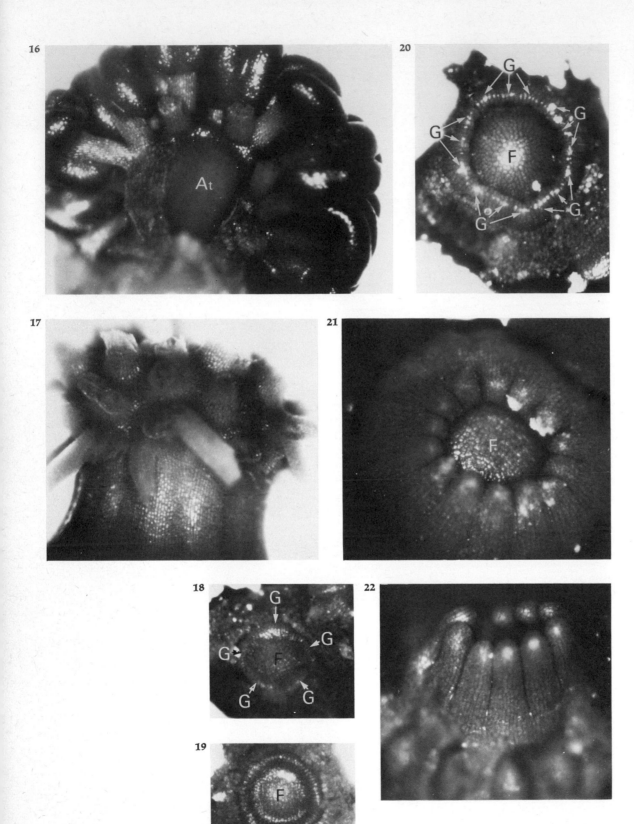

little projections of a crenate leaf margin (20). Usually each primary gynoecial primordium forms three secondary gynoecial primordia; thus, 15 secondary gynoecial primordia are formed altogether. There is, however, some variation in the number of secondary gynoecial primordia. With the formation of secondary gynoecial primordia, septa are initiated on the gynoecial ridge and the floral apex (20). These septa alternate with the secondary gynoecial primordia. Thus, locules are formed below the secondary gynoecial primordia equalling their number (20, 23). In each locule only one bitegmic anatropous ovule develops. This ovule is initiated where the base of the locule merges with the central portion of the floral apex (23). Since the floral apex cannot be sharply delimited from the base of the locule, it is a futile question to ask whether the ovules are formed on the floral apex or on the base of the locule. The styles develop from the secondary gynoecial primordia. Consequently, there are as many styles as there are secondary gynoecial primordia, i.e. often 13-15 (21, 22). The ovarial portion of the gynoecium is formed by the developing gynoecial ridge at the base of the secondary gynoecial primordia.

OTHER AUTHORS

In a detailed study of the Malvales, van Heel (1966) describes the development of the perianth and androecium in *Malva* ssp. His results differ in several respects from my observations. Firstly, van Heel mentions that the petal primordia are initiated when the outer androecial primordia become visible. Secondly, he reports that 'stamen primordia become visible upon them [i.e. the first five antepetalous androecial primordia] in two rows at either side of a radial groove.' Antesepalous androecial primordia are not mentioned. Goebel (1884, pp. 303-4) and Payer (1857) come to a similar conclusion with respect to androecial development. With the exception of figure 8, I could find no evidence for these conclusions. Concerning petal formation, Payer claims that the petals are initiated before the androecium. With regard to gynoecial development neither Payer (1857) nor Goebel (1884, pp. 318-19) mention the occurrence of primary gynoecial primordia. Otherwise their results on gynoecial development agree with my observations. For the work of Frank (1876) and Goebel (1886) on *Malva crispa* (which has only 10 stamens), see van Heel (1966).

BIBLIOGRAPHY

Goebel, K. 1884. *Vergleichende Entwicklungsgeschichte der Pflanzenorgane.* Berlin: R. Friedländer & Sohn.

Heel, W.A. van, 1966. Morphology of the androecium in Malvales. Blumea *13*: 178-394.

Payer, J.B. 1857. *Traité d'organogénie comparée de la fleur. Texte et Atlas.* Paris: Librairie de Victor Masson.

6 Longitudinally dissected androecial tube (At) showing paired arrangement of bisporangiate stamens at and near its upper rim. x 85

7 Side view of undissected androecial tube from which stamens were removed. x 85

FIGURES 18-23. Stages of gynoecial development. x 146

8 Five primary gynoecial primordia (G) with some growth between them. F — floral apex.

9 Secondary gynoecial primordia are not yet clearly discernible.

0. Inception of three secondary gynoecial primordia (G) on each primary one and formation of septa between secondary gynoecial primordia.

2 Further development of secondary gynoecial primordia into styles. Figure 21 is a top view; figure 22 a side view.

3 Top view of young gynoecium from which most of the ovary wall and the septa were removed to show insertion of ovule primordia one of which is labelled (O).

5 Mature flowers. On the flower in figure 25 two petals were removed to show the androecial tube. x ca. 7

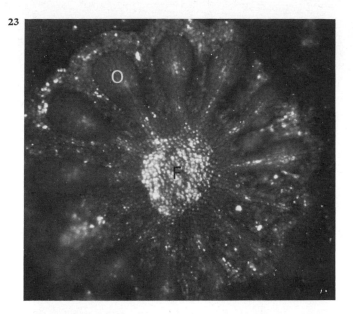

23

24

25

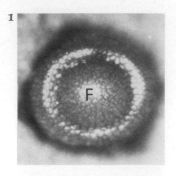

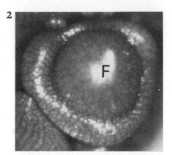

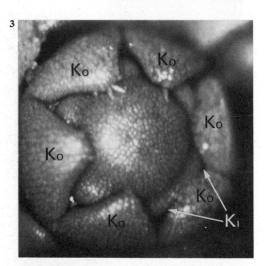

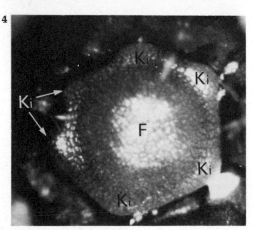

25 / MALVALES (COLUMNIFERAE)

MALVACEAE

Alcaea rosea L.
(= **Althaea rosea** (L.) Cav.)
(hollyhock)

Floral diagram
Note: The aestivation of the corolla varies, as does the number of stamens and styles. Fewer are drawn than actually occur.

Floral formula: $* K_o(6) K_i(5) [C_5 A(\infty)] \underline{G}(\infty) O\infty$

Sequence of primordial inception
$K_o 1\text{-}6$, $K_i 1\text{-}6$, $C 1\text{-}5$ $A 1\text{-}5$, doubling of previous A's, A many in centrifugal sequence and doubling of each overlapping with $G_{girdling}$, $G 1\text{-}\infty$, $O 1\text{-}\infty$
Note: The sepal primordia of each whorl are formed in a very rapid succession which at least in some buds approaches simultaneity.

DESCRIPTION OF FLORAL ORGANOGENESIS

The primordia of the outer sepals (epicalyx members) appear in an extremely rapid succession which approaches simultaneity. During and immediately after their inception, interprimordial growth almost equals the growth of the primordia. Consequently, the initiating outer calyx approaches the shape of a girdling primordium (1). During further development, however, interprimordial growth ceases almost completely, and the primordia of the six outer sepals

become very distinct (2, 3). The primordia of the inner sepals (K_i) appear alternately to those of the outer sepals. They are also initiated in a very rapid sequence that approaches simultaneity (3, 4). In the early stages, interprimordial growth is never as pronounced as in the initiating outer calyx. However, it lasts longer, and therefore produces a conspicuous calyx tube. A peculiarity is the developmental transformation of the originally hexamerous calyx into a pentamerous one: immediately after inception of the primordia, growth occurs between two of the primordia in such a way that they merge into one. One might also say that one of the two primordia merges with the other one as it ceases to grow. Hence, from an early stage onward, only five inner sepal primordia are visible instead of the original six (4, 6, 7). Accordingly, in the mature flower one might expect four of the inner sepals to be alternating with outer sepals and one of the inner sepals (which is formed from two primordia) to be opposite one outer sepal. This is not true. As indicated in the floral diagram, there is equal spacing between all inner sepals and they are of approximately the same size. Thus, the arrangement of sepals in the mature flower does not reflect the peculiarity of their origin.

After inception of the inner calyx, the five petal primordia and the first five antepetalous primordia of the androecium appear. The petal primordia grow extremely slowly during early development. Therefore, it is difficult to decide whether they are initiated immediately before the androecium, after the androecium, or simultaneously with the androecium. In later stages of development, the petal primordia grow rapidly and become slightly two-lobed (27).

To avoid confusion, the first five antepetalous androecial primordia are called primary androecial primordia. As they appear, growth occurs between them. This interprimordial growth is so pronounced that almost a girdling androecial primordium is formed (5-7). Then each of the original five androecial primordia, including the underlying portions of the androecial rim, becomes two-lobed. In other words, five pairs of secondary androecial primordia form in antepetalous positions on the young androecium (8-10). Next, tertiary androecial primordia are initiated in centrifugal sequence on the five pairs of secondary primordia (11-14). The tertiary androecial primordia divide laterally; in this fashion each tertiary primordium forms two quaternary primordia (14-15). Each of these quaternary primordia develops into a filament and a monothecate anther with two pollen sacs (18-25). The two pollen sacs originate very early by a division of each quaternary androecial primordium into

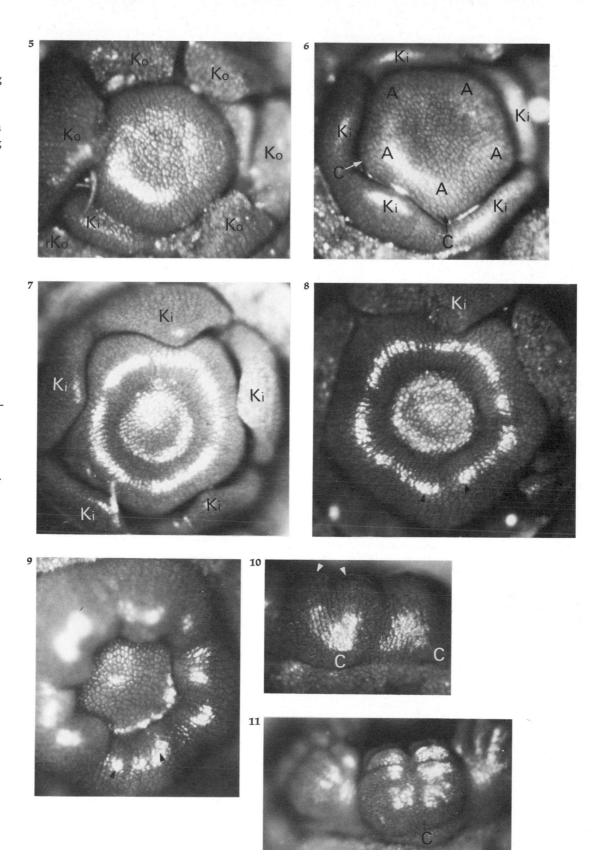

Top views of floral apices showing two stages in the inception of the outer calyx (epicalyx). In figure 2 three of the six outer sepal primordia are quite distinct. Γ = floral apex. x 146

Top views of floral buds showing inception of inner sepals (Ki). In figure 4 the primordia of the outer sepals (Ko) were removed. Arrows point to two inner sepal primordia (Ki) which together will form only one sepal. x 146

Floral buds after partial (5) or complete removal (6) of outer sepals showing inception of petals (C) and primary antepetalous androecial primordia (Λ). x 146

Floral buds immediately before (or during) (7) and after inception (8, 9) of the secondary androecial primordia one pair of which is marked with arrowheads. In figures 8 and 9 the primordia of the inner sepals were removed. The bud of figure 9 was photographed slightly obliquely; therefore one half of the bud is out of focus. x 146

Abaxial view of pairs of secondary androecial primordia before (10) and after (11) inception of tertiary androecial primordia on their outer side. Note that one pair of secondary androecial primordia (arrowheads) is opposite one petal primordium (C) which is still rather indistinct in figure 10. In figure 11 each secondary primordium has only one row of tertiary primordia. x 146

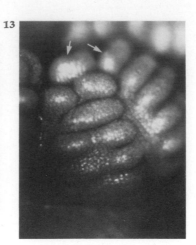

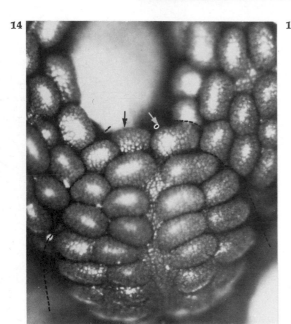

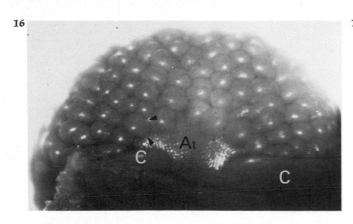

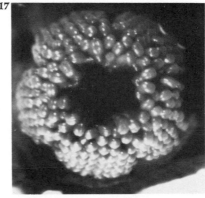

two primordia, each of which then differentiates into a pollen sac (16).

Considerable variation occurs in the formation of the tertiary androecial primordia, and, to some extent, in the subsequent stages of androecial development. In all cases observed, the first tertiary primordium is initiated in a slightly adaxial position near the summit of the secondary primordium. It is followed by a second tertiary primordium, the second one by a third one, etc. This pattern of origin, if continued, would give rise to a row of centrifugally arising tertiary primordia (11). However, instead of any one of these primordia, a pair may occur (12-14). In cases where many of such pairs of primordia are formed, a double row of tertiary primordia is found for at least some distance on the abaxial side of the secondary androecial primordium. For example, on the secondary primordium to the right in figure 14, first two elongate tertiary primordia have been formed at different levels, followed by pairs of tertiary primordia; in this case there is first a single row consisting of two primordia, and then a double row consisting of a larger number of primordia. It can be seen in this case as well as in others, that a double row may be formed in at least two different ways: firstly, as described above, by pairs of primordia occurring at one level; secondly, by alternation of primordia (see, e.g., the lower portion in figure 14 to the right, and figure 12). The two patterns may, however, intergrade with each other. A small amount of growth may occur between adjacent tertiary primordia, especially the two primordia of one pair, i.e., primordia at one level (see, for example, in figure 13 at the lower left side). One cannot exclude the possibility that the interprimordial growth indicates an original common primordium on which the pair of primordia was formed. This would mean that an elongate tertiary primordium produced a pair of quaternary primordia, and consequently the primordia that finally differentiate into pollen sacs would be hexenary instead of pentanary. Several of the pairs of primordia in figures 11-14 could be interpreted in this way. It seems that a final decision on this matter can be made only after observing the development of living floral buds, i.e., tracing the origin of pairs of primordia individually.

In summary, one can state that in the formation of tertiary androecial primordia variation occurs with regard to the number, size, and arrangement of primordia, and the

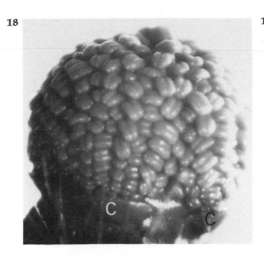

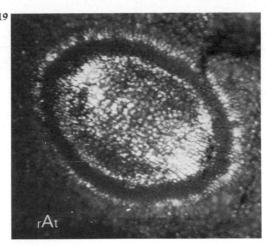

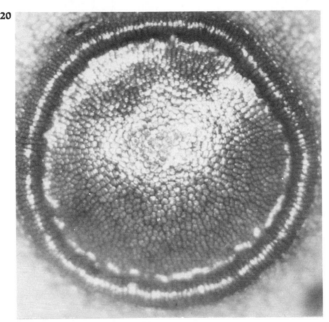

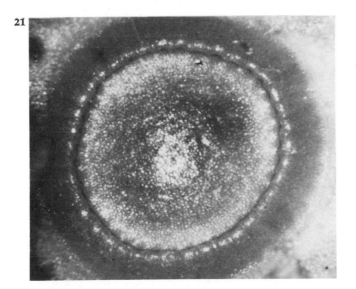

2 Top view of a floral bud from which the young calyx was removed to show the inception of tertiary androecial primordia on five pairs of secondary androecial primordia. In this bud there are two rows of tertiary primordia on all secondary primordia. Some of the tertiary primordia are alternating, others form pairs. x 85

4 Abaxial views of one antepetalous pair of secondary primordia of the young androecium showing different arrangements of tertiary primordia. Arrows point to the first formed tertiary primordia on these and some others show inception of quaternary androecial primordia (= primordia of monothecate stamens). In figure 14 stippled lines delimit one antepetalous androecial complex from two adjacent ones. x 146

5 Side views of floral buds from which the young calyx was removed to show the formation of quaternary (15) and pentanary androecial primordia. Two of the quaternary primordia are marked with arrows (15), and two of the pentanary ones with arrowheads (16). At = primordial androecial tube. x 85

Top view of young androecium at a slightly more advanced stage than that of figure 16. x 37

Side view of much older androecium showing distinct anthers which have become twisted in various directions. The petal primordia (3) are still quite small at this stage. x 37

FIGURES 19-22. Top views of stages in gynoecial development.

Girdling gynoecial primordium forming gynoecial primordia (20) which will develop into styles. The bud of figure 19 was at about the same developmental stage as that of figure 12 (a gynoecial rim may be faintly visible in figure 12). rAt = primordial androecial tube removed. x 146
Formation of locules. x 85

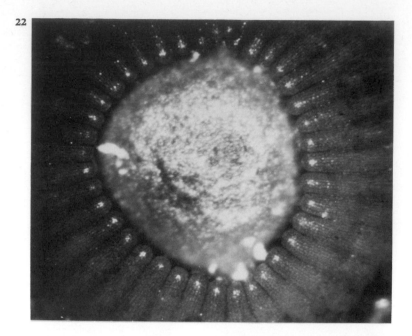

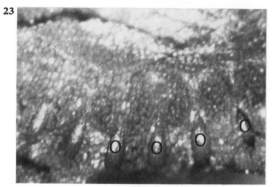

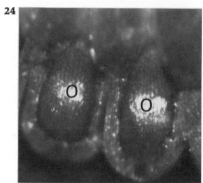

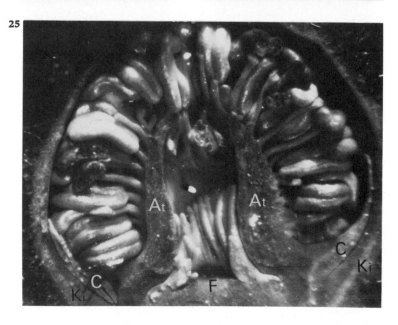

distribution and extent of interprimordial growth. The variation in interprimordial growth may turn out to involve additional branching as mentioned above. In the development of higher order primordia some variation may occur (see Heel 1966); it has not been studied in detail. One would assume that usually the number of monothecate stamens is twice the number of tertiary androecial primordia. Most of the elongation of the androecial tube occurs in later stages of floral development, after the inception of pollen sac primordia (15-18, 25). A slight amount of growth occurs also in a ring-zone below the insertion of the petal primordia. Consequently, the petals become inserted at the base of the androecial tube. The greatest amount of growth occurs in a zone above the insertion of the petal primordia and below the insertion of the lowermost stamen primordia (27).

During the formation of the tertiary androecial primordia, the gynoecium is initiated as a rim surrounding a large floral apex (19). Immediately after its inception, the rim becomes undulated, i.e., many gynoecial primordia appear on the original girdling gynoecial primordium (20). Each of these gynoecial primordia develops into a style (21-22, 25-28). Concomitantly, septa develop from the gynoecial ridge and the adjacent portion of the floral apex in positions alternating with the style primordia (21). Thus, a locule is formed underneath each style primordium. One ovule primordium is initiated at the base of each locule (23). It forms two integuments and becomes anatropous (24). As the gynoecium matures, the wall surrounding each locule bulges outward considerably (24, 26). A large residual apex remains central to the gynoecium.

OTHER AUTHORS

Van Heel (1966) describes briefly androecial development of this species. His results agree with my observation with three exceptions. With regard to the pairs of tertiary androecial primordia, he mentions that the lateral primordia of these pairs arise later. I did not notice such sequential appearance of paired primordia. Furthermore, van Heel reports that the petals become visible 'at a time when the stamen primordia start forming thecae' (p. 201). I observed petal primordia long before (see, e.g., figure 11). However, it may be debated whether they form acropetally before the inception of the primary androecial primordia. According to van Heel, the pistil develops after the formation of thecae. Again, I observed pistil inception at a much earlier stage of development.

In addition to normal flowers, van Heel describes and illustrates the extremely complicated development of 'double flowers' (p. 203). He also quotes and reviews briefly the older literature on the development of normal and double flowers in this species (Goebel 1886; Goethart 1890; Eichler 1875; Sachs 1874). Payer (1857) does not study this species; however, he describes in great detail the floral organogenesis of related genera.

Sankhla *et al.* (1966) reports on the *in vitro* production of 'double flowers' and transformation of anthers into petals. They also find that flowers with a polyadelphous androecium occurred in their cultures.

BIBLIOGRAPHY

Heel, W. A. van. 1966. Morphology of the androceium in Malvales. Blumea 13: 178-394.
Payer, J.B. 1857. *Traité d'organogénie comparée de la fleur. Texte et Atlas.* Paris: Librairie de Victor Masson.
Sankhla, N., Daksha Baxi, and Chatterji, U.N. 1966. *In vitro* production of petals from anthers of *Althaea rosea* (Hollyhock). Curr. Sci. 35: 574-575.

2 Elongation of styles. x 85
4 Top view of portion of the young gynoecium after removal of most of the gynoecial wall and septa to show ovule primordia (O) in a young (23) and older developmental stage (24). Figure 23, x 146. Figure 24, x 85
5 Longitudinally dissected floral bud showing residual floral apex (F), many primordia of styles (unlabelled), young androecial tube (At) with stamens, young petals (C), and inner sepals (K₁). x 18
6 Gynoecium of a rather old floral bud. x 18
7 Mature flowers in top view (28) and side view after removal of two petals (27). x ca. 1

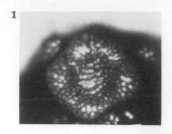

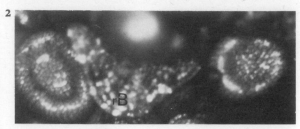

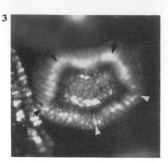

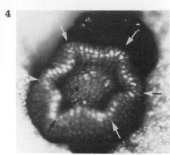

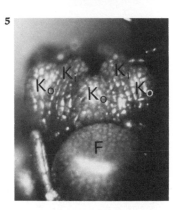

26 / MYRTIFLORAE

LYTHRACEAE

Lythrum salicaria L.
(purple loosestrife)

Floral diagram
Note: The orientation of the septum varies; on the same inflorescence the septum was found in the median and transversal plane.

Floral formula: $\ast$ [K6 + 6 C6 A6 + 6] $\underline{G}$(2) O∞
Note: Normally the flowers are hexamerous, but pentamerous flowers are not rare. The species is characterized by trimorphic heterostyly.

Sequence of primordial inception
K_i1-6, K_o1-6, A_o1-6, $G_{girdling}$, A_i1-6, C1-6, O many in basipetal sequence
Note: The members of each whorl tend to be initiated in a very rapid succession which approaches simultaneity. At least in some buds, it is more appropriate to write G1-2 instead of $G_{girdling}$.

DESCRIPTION OF FLORAL ORGANOGENESIS

First, the primordia of the inner sepals are initiated in a very rapid sequence which approaches simultaneity (1). Because of this rapidity in origin, the sequential pattern is difficult to determine. Furthermore, hardly tangible variation occurs in the questionable sequence (1-3, 6). Interprimordial growth leads to the formation of a primordial calyx tube in a very early stage of development (1-4). In some floral buds interprimordial growth is so pronounced in the very early stages that the incipient calyx approaches the shape of a girdling primordium. The primordia of the outer sepals are initiated at the upper rim of the primordial calyx tube (3-5). Whereas the primordia of the inner sepals grow more or less horizontally toward the center of the bud, the primordia of the outer sepals and the calyx tube grow in a vertical direction (5).

After calyx inception, the primordia of the outer stamens are initiated at the base of the primordial calyx tube opposite the inner sepal primordia (4, 6). Next, the floral apex gives rise to a girdling gynoecial primordium. The outline of this primordium varies: it may be round, oval, or somewhat

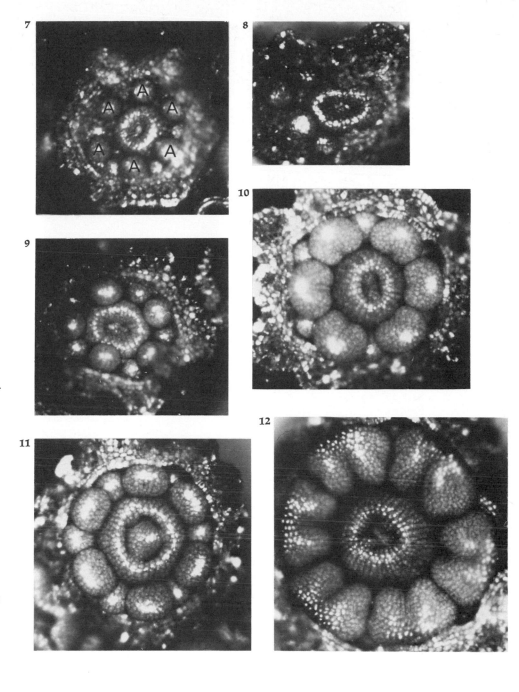

4 Top views of floral buds showing inception of inner sepals and calyx tube. In figures 3 and 4 arrows point to initiating outer sepal primordia. Some of the primordia of outer stamens are perhaps faintly visible in figure 4. The buds of figures 1 and 3 are pentamerous. Figure 2 shows two buds and the base of a bract that was removed (rB). x 146

5 Floral bud showing young calyx tube with primordia of inner (K_1) and outer (K_0) sepals. At the bottom of the photograph is another floral apex (F) showing calyx inception. x 146

6 Top view of three floral buds: the two lateral ones showing inception of inner sepals; the central one showing gynoecial inception; the young calyx tube and one of the six outer stamen primordia (A) were removed (rA). x 146

8 Top views of two dissected floral buds with incipient gynoecia of various shapes. In figure 7 only the outer stamen primordia are labelled (A). In figure 8 some of the stamen primordia were removed. x 146

2 Top views of floral buds from which the young calyx was removed to show the primordia of the larger outer stamens, the smaller inner stamens, and the pistil wall and septa (10-12). In figure 12 the smaller inner stamens are not visible. Figure 11 shows an unusual bud with a young trimerous gynoecium. x 146

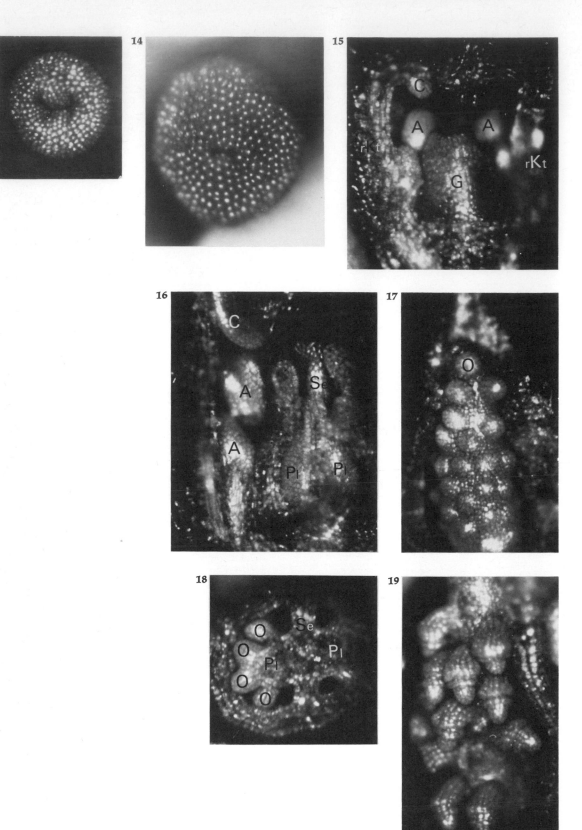

irregular (6-9). Besides that, it may be slightly two-lobed (6). In this case it is, of course, more appropriate to say that the gynoecium is formed by two gynoecial primordia which become interconnected by interprimordial growth immediately after their inception. As the gynoecium grows upward, the two indistinct gynoecial primordia (or two lobes of the girdling gynoecial primordium) do not develop further. Consequently, older developmental stages of the gynoecium are tubular (10-13). Two septa (rarely three as in figure 11) grow out from opposing sides of the primordial pistil wall. They are continuous across the floral apex (10-12). As a result, the ovarial portion of the pistil becomes subdivided in two locules by one solid upgrowth, whereas the stylar and stigmatic portions are tubular with the two septa meeting in the centre (13). Complete postgenital fusion finally produces a stigma that does not show the slightest indication of its origin (figure 13 still shows a trace of the original opening). The solid upgrowth (septum) in the center of the young ovary grows out on both sides, thus producing two placentae (16). Many ovules are initiated in a basipetal sequence (17). Each ovule becomes bitegmic and anatropous (19).

Immediately after gynoecial inception, the primordia of the inner stamens are initiated (9). Like those of the outer stamens, they develop into stamens with introrse anthers.

Finally, the petal primordia are initiated high up on the primordial calyx tube (15). As they develop, they grow downward and become profusely twisted and folded.

Some growth occurs in a zone below the stamen primordia on the young calyx tube. This leads to the insertion of the stamens at the base of the calyx tube.

Three types of flowers occur on different plants as a result of differential elongation of outer stamens, inner stamens, and style (20-22).

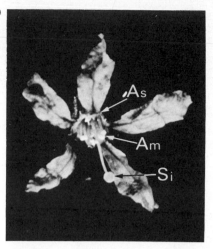

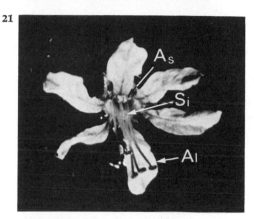

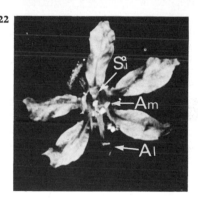

Cheung and Sattler (1967) describe the early floral development of this species. The present work was done with material from the same colony as that used for the earlier study. Nonetheless, more variation was found than reported previously: calyx inception appears to be rather plastic and the same is true for the initiation of the gynoecium. Otherwise, this work confirms the previous report. However, a comparison of magnifications seems to indicate that an error must have been made by Cheung and Sattler (1967) in calculating the magnifications as they appear to be too high.

Mayr (1969) describes the early floral development of *Lythrum virgatum* L. Her observations agree with ours on *Lythrum salicaria* L.

Stout (1925) reports a fourth form of flowers in which 'the pistil and the set of longer stamens are almost of equal length and this length is about midway between the two lengths characteristic of the mid- and the long-lengths as seen in long- and in mid-styled plants' (pp. 81-82).

Schoute (1932), using only mature flowers, studies the variation in number and position of floral parts, including vascularization, in over 1500 flowers.

BIBLIOGRAPHY

Cheung, M. and Sattler, R. 1967. Early floral development of *Lythrum salicaria*. Can. J. Bot. *45*: 1609-1618.

Mayr, Barbara. 1969. Ontogenetische Studien an Myrtales-Blüten. Bot. Jb. *89*: 210-271.

Schoute, J.C. 1932. On pleiomery and meiomery in the flower. Rec. trav. bot. néerl. *29*: 164-226.

Stout, A.B. 1925. Studies of *Lythrum salicaria*. A new form of flower in this species. Bull. Torrey Bot. Club *52*: 81-85.

13 Top view of young pistil showing the primordial septa. x 146

14 Top view of young stigma shortly before complete postgenital fusion. x 146

15 Side view of floral bud showing young pistil wall (G), young stamens (A), and petal primordium (C) high up on the young calyx tube (Kt). A portion of the young calyx tube was removed. x 146

16 Side view of longitudinally dissected bud showing one septum (Se), both placentae (Pl) before ovule inception, two young stamens (A), and a petal primordium (C). x 146

17 Side view of a dissected young ovary showing basipetal ovule inception on one placenta. This view is at right angles to the one in figure 16. Only the uppermost ovule primordium is labelled (O). x 146

18 Cross-section through part of young ovary showing the partition (Se) and one of the two placentae (Pl) with ovule primordia (O). x 146

19 Part of a placenta with ovule primordia showing inception of integuments. x 146

FIGURES 20-22. Mature flowers. Two of the three flowers are pentamerous (20, 22); one is hexamerous (21). Si = stigma, As = short stamens, Am = medium stamens, Al = long stamens. x ca. 3

20 Long-styled flower.

21 Medium-styled flower.

22 Short-styled flower.

27 / MYRTIFLORAE

ONAGRACEAE

Fuchsia hybrida Voss
(= **F. speciosa** Hort.)
(garden fuchsia)

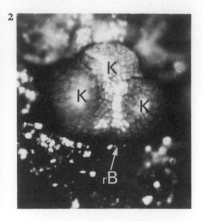

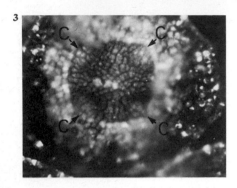

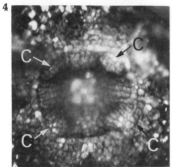

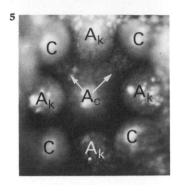

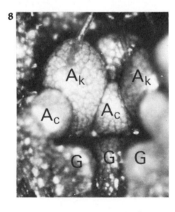

Floral diagram
Note: The disc is not indicated in the floral diagram.

Floral formula: $\ast\ [K(4)\ C_4\ A_4 + 4\ D]\ \overline{G}(4)\ O\infty$

Sequence of primordial inception
$K_{1-2,3-4}$, C_{1-4}, A_k1-4, A_c1-4, G_{1-4}, $O_{1-\infty}$, D

DESCRIPTION OF FLORAL ORGANOGENESIS

On the floral apex, first the two transversal sepal primordia are initiated. Then the two median sepal primordia appear (1). In at least some buds the adaxial sepal primordium is formed before the abaxial one (2). Moreover, a slight difference in size of the two transversal sepal primordia might indicate that these also appear successively (2). Consequently, the four sepal primordia are initiated successively at least in some buds (2). Immediately after sepal inception growth extends from the sepal primordia to the areas between their margins. This leads to the formation of a calyx tube even before the inception of any further appendages.

The four petal primordia are initiated at about the same time high up on the primordial calyx tube (3). Then the primordia of the antesepalous stamens are formed slightly below (centripetal to) the petal primordia (5). When they have reached a considerable size, the primordia of the antepetalous stamens appear in the areas between them. With regard to their outer margins, the antepetalous stamen primordia are located inside the antesepalous ones. This means that the developing androecium is diplostemonous in terms of the outer margins of the stamen primordia. Whether

9

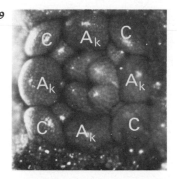

12

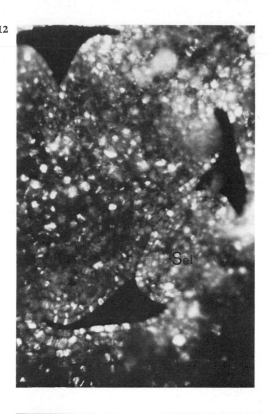

10

11

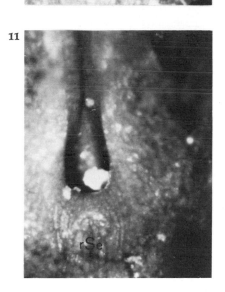

13

1 Top view of inflorescence (shoot) tip showing two floral buds with large transversal sepal primordia (K) and incipient median sepal primordia (arrowheads). The subtending bract primordia were removed (rB). Two smaller bract primordia (B) are present at a higher level. x 146

2 Top view of a floral bud showing unequal size of the four sepal primordia. The abaxial one (arrowhead) is the smallest. x 146

5 Top views of floral buds from which the sepal primordia were removed to show inception of petals (3, C), antesepalous stamens and perhaps antepetalous stamens (5, A$_K$ A$_c$). x 146

6 Incipient gynoecial primordia (arrowheads). x 146

7 Floral bud from which primordia of sepals, petals and antesepalous stamens (rA$_K$) were removed to show gynoecial inception. x 146

8 Longitudinally dissected bud of a developmental stage like that of figure 7. Three of the gynoecial primordia (G) are visible. x 146

9 Top view of a floral bud from which sepal primordia were removed. The gynoecial primordia and the primordia of the antepetalous stamens are unlabelled. x 85

10 Longitudinally dissected young ovary showing origin of the first septum. x 146

11 Interior of ovary viewed perpendicularly to that of figure 10. The second septum was removed in front (rSe). Note that the first one grows up to a higher level. x 146

12 Cross-section through young ovary at a level where the first septum is still solid in the center, but the second septum present only by its lateral portions (Sel) one of which is fully shown. x 146

3 Cross-section through a young ovary at a level where the four lateral portions of both septa fuse postgenitally in the center (see arrowheads). x 146

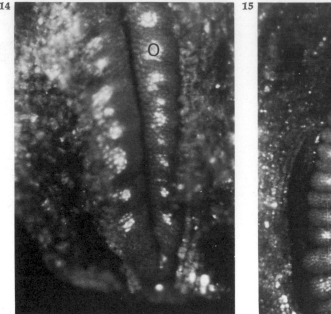

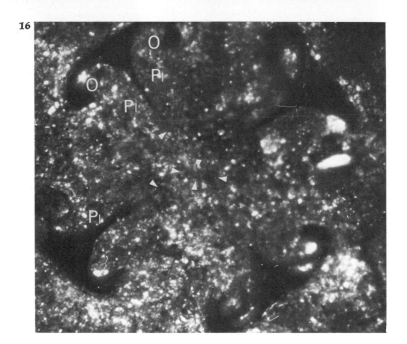

the same is true for the inner margins of these primordia is doubtful. Each stamen develops a filament and an introrse anther.

After androecial inception, four gynoecial primordia are initiated inside the antepetalous stamen primordia on the young calyx tube (6-8). Interprimordial growth occurs immediately after their inception; thus, a cylindrical portion grows upward with the four gynoecial primordia on top of it (9). The cylindrical portion develops into a long style, and the four gynoecial primordia produce four stigmas. Concomitantly, the ovary develops due to upgrowth in a region below the insertion of the gynoecial cylinder. The two septa are formed successively; they arise by growth along the ovarial wall and the floral apex (10). One of them grows to a slightly higher level than the other (11, 12). In the upper region of the ovary the lateral portions of both septa fuse postgenitally in the center of the ovary (13). Thus, three regions may be distinguished with regard to the formation of the four locules: in the lower region, solid upgrowth of two septa from the floor of the ovary and in continuity with the ovarial wall; in a small region above, further solid upgrowth of one of the septa and ingrowth of the lateral portions of the other septum (12); and, in the upper region, ingrowth of the lateral portions of both septa (13). Two placentae are formed in each locule throughout the three regions. One row of ovule primordia appears on each placenta (14-16). Each ovule primordium forms two integuments and assumes an anatropous condition.

During ovarial development, upgrowth occurs in a zone below the insertion of the young stamens and above the insertion of the gynoecial cylinder (style). This leads to the formation of what is usually called a hypanthium. One might also call it a calyx tube upon whose upper margin the petals and stamens are inserted. Regardless of terminology, the developmental pattern is the following. After the inception of sepal primordia, a tube is formed at their base. With the initiation of the gynoecial cylinder, this tube has a portion above and a portion below the insertion of the gynoecial cylinder. The portion below develops into the ovary wall, the portion above into the hypanthium (or calyx tube).

While the hypanthium (or calyx tube) grows upward, a disc is initiated at its base in the form of a circular primordium (17). As it matures, this primordium forms four inconspicuous two-lobed bulges opposite the petals (18).

OTHER AUTHORS

I found no detailed literature on floral development of *Fuchsia*. Mayr's (1969) brief description of the early floral development of *Fuchsia gracilis* Lindl. agrees with my observations on *Fuchsia hybrida*. Kaienburg (1950-51; quoted by Kowalewicz 1956) studies mainly older stages and mature flowers, and Baehni and Bonner (1949) investigate the vascularization of the hypanthium.

Related genera of the same family have been investigated in much more detail (see, e.g., Pankow 1966; Kowalewicz 1956; Barcianu 1875; Payer 1857).

BIBLIOGRAPHY

Baehni, Ch. et Bonner, C.E.B. 1949. La vascularization du tube floral chez les Onagracées. Candollea 12: 345-359.
Barcianu, D.P. 1875. Untersuchungen über die Blüthenentwickelung der Onagraceen. Schenk u. Luerssen's Bot. Mitth. 2: 81-129.
Kowalewicz, R. 1956. Entwicklungsgeschichtliche Studien an normalen und cruciaten Blüten von *Epilobium* und *Oenothera*. Planta 46: 569-603.
Mayr, Barbara. 1969. Ontogenetische Studien an Myrtales-Blüten. Bot. Jb. 89: 210-271.
Pankow, H. 1966. Histogenetische Untersuchungen an den Blüten einiger *Oenothera* – Arten. Flora, Abt. B, 156: 122-132.
Payer, J.B. 1857. *Traité d'organogénie comparée de la fleur. Texte et Atlas.* Paris: Librairie de Victor Masson.

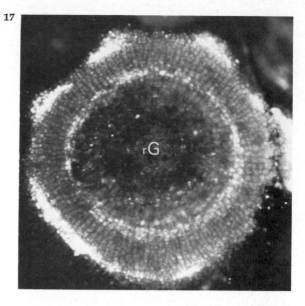

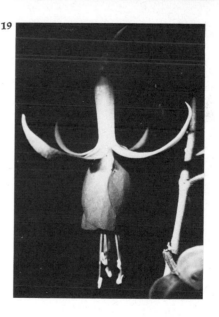

Side view of placentae of adjacent septa showing ovule inception. Only one ovule is labelled (O). x 146
Side view of one placenta with a row of ovule primordia. x 146
Cross-section through upper portion of the ovary showing the four lateral portions of the two septa with the placentae (Pl) and ovule primordia (O) on either side. Arrowheads point to the postgenital fusion line of the septal portions. x 85
Top view of primordial disc. The upper (free) portion of the young pistil was removed in the center (rG). x 85
Portion of the young disc. The two lobes are opposite the petal which is not shown. x 85
Mature flower. x ca. 1

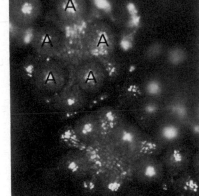

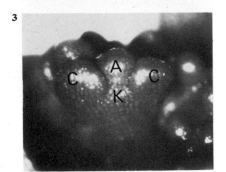

28 / UMBELLIFLORAE

UMBELLIFERAE (APIACEAE)

Anthriscus sylvestris (L.) Hoffm. (wild parsley)

Floral diagram

Floral formula: $*C5 \, A5 \, \overline{G}(2) \, O2$
Note: No distinct calyx is present. But rudiments which may be interpreted as a vestigial calyx are faintly visible. In addition to the two functional ovules two rudimentary ovules are present (see floral diagram).

Sequence of primordial inception
$A1\text{-}3,4\text{-}5 \, C1\text{-}5, \, G1\text{-}2, \, K_{rud}.1\text{-}5, \, O1\text{-}4$
Note: The first petal primordia may appear before the last two stamen primordia. All stamen and petal primordia may be initiated in a very rapid succession.

DESCRIPTION OF FLORAL ORGANOGENESIS

First, two stamen primordia toward the abaxial side of the floral apex and the adaxial stamen primordium are initiated at about the same time (1). Possibly their appearance is not simultaneous; one of the primordia on the abaxial side may appear first, followed by the adaxial one and then the other one on the abaxial side. The two lateral stamen primordia originate after the first three stamen primordia have become

rather prominent. They are formed at about the same time; however, again it is possible that one of them precedes the other. Consequently, there is a possibility that the five stamen primordia appear in a spiral sequence, as indicated in the floral diagram. Even so, the plastochron between the third and fourth primordia is much greater than the plastochron between the first three primordia on the one hand, and the last two primordia on the other hand.

When the last two stamen primordia appear, one, or even more than one, of the petal primordia is already visible. Probably, the first petal primordia arise simultaneously with the last two stamen primordia; or perhaps the last two stamen primordia are formed after one, or more than one, petal primordium has been initiated. With respect to each other, the petal primordia are formed at about the same time. However, again there is a possibility that they appear in a spiral sequence with an extremely small plastochron. If one assumes the latter, then the first petal primordium would be the one alternating with the two stamen primordia toward the abaxial side.

After the inception of all petal primordia, the young floral bud elongates in median direction (2). While this elongation occurs, the gynoecium appears as two crescent-shaped primordia (4, 5). Immediately after their inception, the two crescent-shaped primordia become horseshoe-shaped (5, 6), thus forming two septa in the transversal plane (the plane of the shorter bud diameter). The two septa meet in the center of the developing gynoecium (9). Each is double at first (6, 7), since it is formed by the margins of the two adjacent gynoecial primordia (carpel primordia). Subsequently, uniform growth below the two double septa gives rise to two solid septa with the original double septa on top of them (7, 10). Zonal growth at the base of the developing corolla, androecium, and gynoecial wall leads to the inferior position of the ovary. Actually, growth occurs in the entire developing receptacle except at the base of the two gynoecial locules. Thereby the locules remain inferior to the other floral parts (9, 14, 15, 16). As the ovary wall grows in diameter, two deep grooves are formed on the same radius but outside the gynoecial septa due to diminishing growth in these two regions (14, 15, white arrowheads). The median part of the two gynoecial primordia grows faster than the septal region and thus

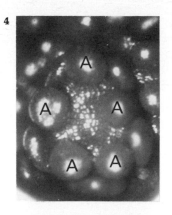
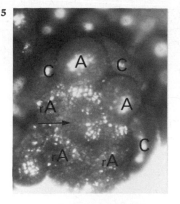
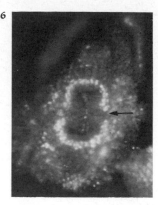

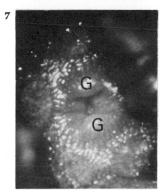
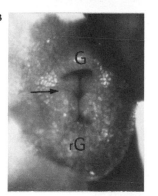

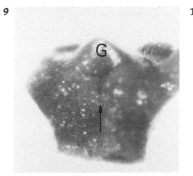
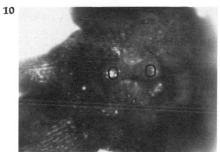

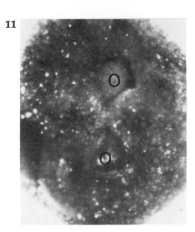

1 Top view of young umbellet with the inflorescence apex (R) in the center, surrounded by floral apices (F) and young floral buds in very early stages of development. At one bud to the left the inception of the two stamen primordia toward the abaxial side is marked with white arrowheads, whereas the inception of the adaxial stamen primordium is marked with a black arrowhead. At the floral bud at the top of the photograph all five stamen (A) and petal (C) primordia have been labelled. The adaxial stamen primordium is the one near R. x 146

2 Two floral buds with stamen primordia (A) which are labelled only in the upper bud; petal primordia are not labelled. x 146

3 Side view of a floral bud at the time of gynoecial inception. Outside the stamen primordium (A) a bulge of tissue is visible (K). x 146

4 Top view of a floral bud showing the inception of the gynoecial primordia. x 146

5 Top view of a floral bud from which three stamen primordia were removed (rA). The gynoecial primordia, on top and bottom in the photograph, are beginning to form the septa (see arrow). x 146

6 Two gynoecial primordia forming the septa (see arrow). x 146

8 The two gynoecial primordia (G) in a more advanced stage of development. The arrow in figure 8 points to the plane of the septa. x 146

9 Longitudinal section of the developing pistil in the plane of the septa. The arrow points to the suture of the septa in the center of the pistil. Epigyny is already marked at this stage of development. x 146

10 Cross-section through the developing ovary. On one septum two ovule primordia (O) are visible. x 146

11 Cross-section through a more advanced ovary with one ovule visible in each locule. x 146

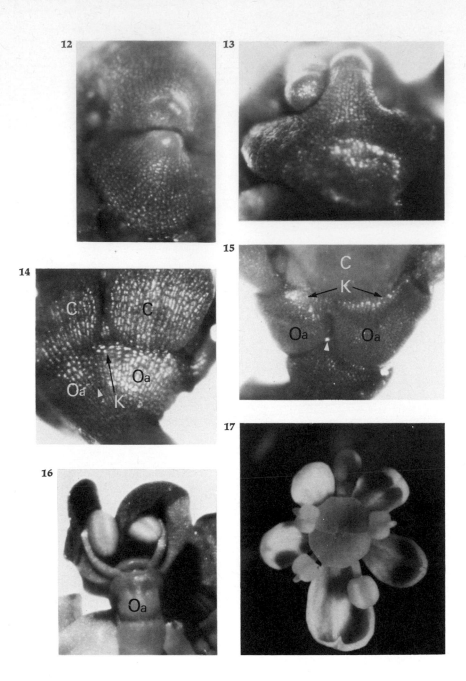

the pistil becomes gradually closed, and the upper part of each primordium forms a style (8, 12, 13).

Two ovule primordia per locule are formed toward the upper part of the septal margin (10). Only one ovule per locule reaches maturity. It is inserted near the top of the ovary, i.e., it is pendent. It has only one integument.

A distinct calyx is not present. But at about the time of gynoecial inception or soon afterwards, an inconspicuous bulge of tissue becomes visible at locations outside and on the same radius as the stamen primordia (3). In later developmental stages growth extends from these (sepal) positions all around the corolla (14, 15, 16), but this growth is so limited that it is questionable whether one should refer to the product as a calyx.

At the periphery of the inflorescence, the flowers become slightly zygomorphic due to unequal growth of the petals (17).

OTHER AUTHORS

Schumann (1890) studies the early floral organogenesis of this species and finds that the petals are initiated immediately after the stamens or perhaps sometimes simultaneously with the stamens. In the peripheral flowers he notes a sequence of stamen inception which deviates from the one reported here, but in some central flowers he observes a sequence as I did. Payer (1857) describes the floral development of *Heracleum barbatum*, and he mentions that other genera of the Umbelliferae show the same floral development. In contrast to my observations (figures 1, 2) he reports a centripetal sequence of inception with regard to all appendages. Jurica (1922) who studies mainly *Eryngium yuccifolium* concludes also that floral development in the Umbelliferae 'shows an acropetal succession of floral cycles' (p. 301). He adds that 'in genera in which the sepals are represented by mere calyx teeth or are obsolete, the calyx primordia also make their appearance, but fail to develop any further' (p. 301).

Sieler (1870) finds three types of floral development in the Umbelliferae. In the first type the sequence is the following: stamens, petals, sepals, carpels. In the second type stamens and petals are formed simultaneously, followed by the sepals and carpels. In the third type the inception of the adaxial petal is followed by the initiation of the two adjacent stamens. Then the two lateral petal primordia appear, followed by the two adjacent stamens. Finally, the two petals toward the abaxial side are initiated, followed by the adaxial stamen. Thus, petals and stamens are formed in a direction from the adaxial side of the floral apex toward the abaxial side. This sequence contrasts with the spiral or simultaneous appearance of primordia. The sepals and carpels are formed after all petal and stamen primordia have been initiated. My observations on *Anthriscus sylvestris* seem to agree with Sieler's first type. However, if the first petals appear before the inception of the lateral stamens, then the pattern of inception may be intermediate between Sieler's first and third types.

Borthwick *et al.* (1931) do not commit themselves on the sequence of primordial inception in *Daucus carota*. They write: 'Sepal, petal, and stamen primordia are formed nearly simultaneously. These are followed by carpel primordia.'

Hakansson (1923) who studied the embryology of Umbelliferae quotes older literature on floral development.

BIBLIOGRAPHY

Borthwick, H.A., Phillips, M., and Robbins, W.W. 1931. Floral development in *Daucus carota*. Amer. J. Bot. *18*: 784-796.

Hakansson, A. 1923. Studien über die Entwicklungsgeschichte der Umbelliferen. Lunds Univ. Arssk. N.F. II, *18*: 1-120.

Jurica, H.S. 1922. A morphological study of the Umbelliferae. Bot. Gaz. *74*: 292-301.

Payer, J.B. 1857. *Traité d'organogénie comparée de la fleur. Texte et Atlas.* Paris: Librairie de Victor Masson.

Schumann, K. 1890. *Neue Untersuchungen über den Blüthenanschluss.* Leipzig.

Sieler, T. 1870. Beiträge zur Entwicklungsgeschichte des Blütenstandes und der Blüte bei den Umbelliferen. Bot. Zeit. *28*: 361-369, 377-382.

---

13 Developmental stages of the two styles with a stylopodium at their base. x 146

16 Side views of two developing and one mature (16) flower. The white arrowhead points to the developing groove on the inferior ovary (Oa). The petals (C) and the rudimentary calyx (K) are visible. On the flower in figure 16 one petal between the two visible stamens was removed. Figure 14, x 146. Figure 15, x 85. Figure 16, x ca. 28.

17 Top view of a mature flower from the periphery of the inflorescence showing slight zygomorphy. x ca. 28

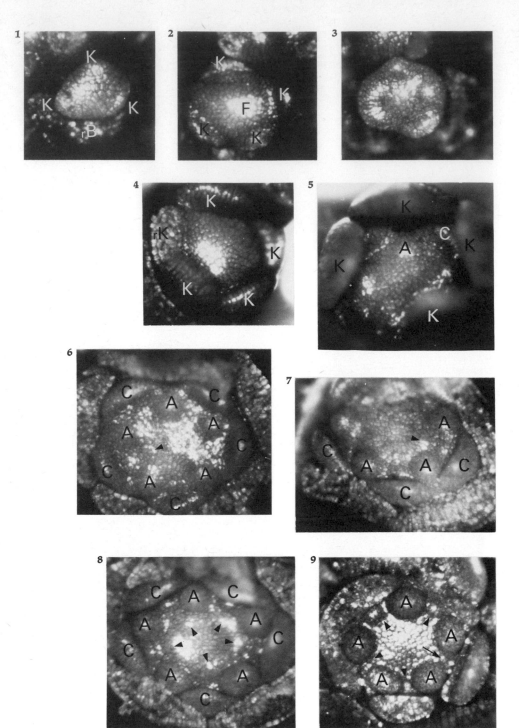

29 / ERICALES

PYROLACEAE
Pyrola elliptica Nutt. (shin-leaf)

Floral diagram
Note: The aestivation of the corolla is very variable. The pattern in the floral diagram is only one among several others that occur.

Floral formula: $*$K(5) C5 A5 + 5 $\underline{G}$(5) O∞
Note: When the flower opens, it becomes slightly zygomorphic due to the orientation of the style and stamens.

Sequence of primordial inception
K1,2,3,4,5, C1-5, A_k1-5, A_c1-5, G1-5, O1-∞

DESCRIPTION OF FLORAL ORGANOGENESIS

The five sepal primordia appear in a very rapid spiral sequence (1, 2). In a number of floral buds the first three or four sepal primordia are about the same size. This may indicate that they have been initiated at about the same time. In these cases, a genetic spiral of primordial inception could not be visualized (3). Growth between the margins of the sepal primordia starts before the sepal primordia overarch the floral apex (4).

The five petal primordia are initiated at about the same time (4). They are slightly crescent-shaped when the antesepalous stamens appear simultaneously (5, 6). The antepetalous stamens are initiated after the antesepalous ones and at about the same level with regard to the center and outer (abaxial) margin of the primordium (7-9). If one considers the inner (adaxial) margin of the stamen primordia, the antepetalous stamen primordia are clearly outside the antesepalous ones, i.e. only with regard to the inner limits of the stamen primordia is the inception of the androecium obdiplostemonous. During the further development of the flower, the antesepalous stamens

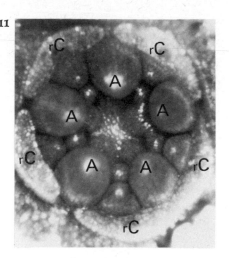

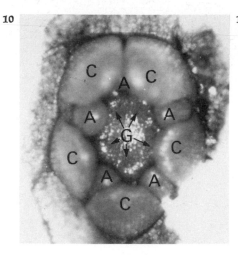

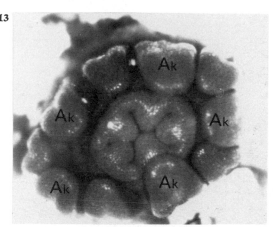

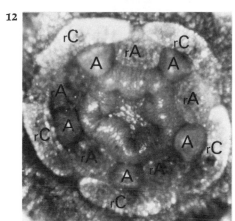

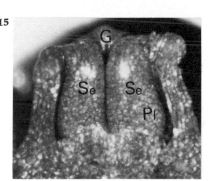

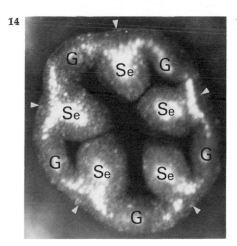

-3 Top views of floral apices showing sepal inception. In figures 1 and 2 a genetic spiral can be visualized, not in figure 3. x 146

-5 Slightly oblique top views of floral buds. Figure 4 shows petal inception alternately to the sepal primordia (K) of which one was removed (rK). Figure 5 shows inception of antesepalous stamens, one of which is labelled (A). x 146

-8 Floral buds from which sepal primordia were removed to show inception of antepetalous stamens and gynoecial primordia. Only the antesepalous stamen primordia are labelled (A). Arrowheads point to some of the gynoecial primordia. x 146

9 Unusual floral bud which probably was arrested in development. (Note that some of the buds near the tip of the inflorescence abort.) The young perianth was removed to show the complete lack of at least one antepetalous stamen primordium (see arrow) and the marked suppression of the others (see arrowheads). x 146

10 Floral bud from which the young sepals were removed to show inception of the pistil wall by five primordia (G). x 146

11 Floral bud from which the young perianth was removed to show inception of the gynoecial septa opposite the antesepalous stamens primordia (A). x 146

12 Same as in figure 11, after removal of antesepalous stamen primordia (rA). x 146

3 Floral bud after removal of the young perianth. A_k = antesepalous stamen primordium. x 146

4 Top view of pistil in a developmental stage like that of figure 13. The gynoecial primordia (G) correspond to those in figure 10. Arrowheads point to incipient growth between the gynoecial primordia. Se = young septum. x 146

5 One half of a young pistil after longitudinal splitting of a whole young pistil showing development of septa (Se) and inception of placenta (Pl). x 146

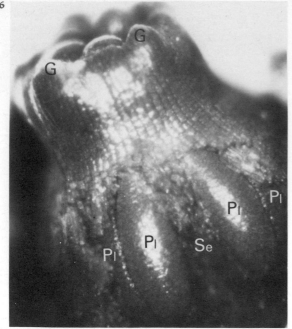

16

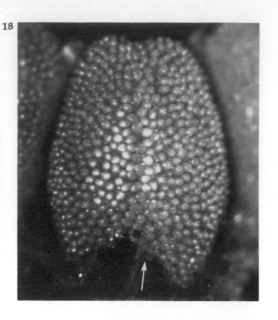

18

19

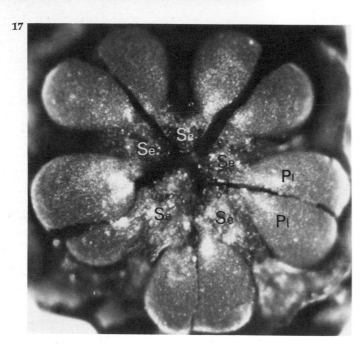

17

grow more toward the center of the flower than the antepetalous ones. Consequently, the androecium of the mature flower is obdiplostemonous even with regard to the center of the stamens. The individual stamens consist of a filament and an introrse anther which approaches the latrorse condition. The tips of the anthers are turned almost 180 degrees towards the adaxial side of the filaments. As a result, the pores which are morphologically at the base of the anther become terminal (22).

The five gynoecial primordia are initiated opposite and immediately after the antepetalous stamens. The possibility that the primordia of the gynoecium and the antepetalous stamens appear simultaneously cannot be excluded with certainty. However, since the primordia of the antepetalous stamens appear slightly larger than those of the gynoecium, it is likely that they are initiated first (6-8). Immediately after the inception of the five gynoecial primordia, growth extends between their margins (10). As a result, a five-lobed cylinder grows up which will develop into the pistil wall. Five septa are initiated very early in antesepalous positions. They are formed on the floral apex as radial extensions of the primordial pistil wall (11-15). When they meet in the center of the young pistil, each of them forms two lateral placentae which grow toward the pistil wall. The placentae of adjacent septa eventually come into close contact and form one compound placenta (16, 17). Many ovules are formed simultaneously on all placentae (18, 19). Each ovule develops one integument and becomes anatropous (19). After ovule formation the line where two placentae of adjacent septa come into contact is no longer visible (18).

As the pistil wall and the septa grow upward, they form an ovary portion and a style (20). Up to the latest stage of development, the septa extend from the base of the ovary to the tip of the style. Eventually their distal surface will become stigmatic (21,

22). The five original lobes of the primordial pistil wall (see figure 10) are inconspicuous in the later stages of development because they have grown very little; but they are still recognizable alternate to the massive tips of the septa (14-16). In these later stages growth extends between the original gynoecial primordia (i.e. the original lobes of the gynoecium) and as a result, a rim is formed around the tips of the septa (14, 16, 20, 21). This means that the gynoecium begins its development by five gynoecial primordia and afterwards growth between the primordia is such that a completely girdling structure (i.e. an unlobed cylindrical wall) results.

OTHER AUTHORS

Leins (1964) describes the floral development of *Pyrola rotundifolia* L. and *Pyrola minor* L. with special reference to the androecium. There seems to be no discrepancy between my observations on *Pyrola elliptica* Nutt. and those of Leins on the other two species. Pyykkö (1969) describes the anatomy of the pistil in other species of this genus. He notes two zones: a synascidial and a synplicate zone both of which comprise about half the length of the ovary. Hunt's (1937) study of the mature style and stigma is largely interpretive and speculative.

BIBLIOGRAPHY

Hunt, K.W. 1937. A study of the style and stigma with reference to the nature of the carpel. Amer. J. Bot. *24*: 288-295.

Leins, P. 1964. Entwicklungsgeschichtliche Studien an Ericales-Blüten. Bot. Jb. *83*: 57-88.

Pyykkö, M. 1969. Placentation in the Ericales. I. Pyrolaceae and Monotropaceae. Ann. Bot. Fenn. *6*: 255-268.

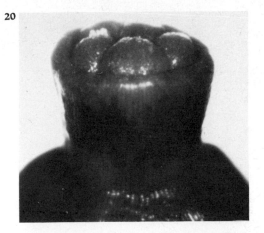

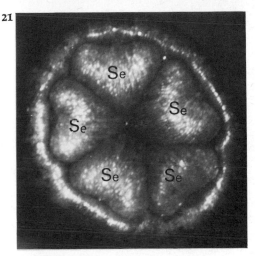

---

6 Side view of young pistil after part of the ovary wall was removed to show placentae before ovule inception. x 146

7 Top view of septa with placentae before ovule inception; pistil wall removed. x 85

8 Side view of two adjacent placentae which cannot be distinguished after the inception of very many ovule primordia. The white arrow points to the obscured contact line of the placentae. x 85

9 Ovules on the placentae of figure 18. x 146

0 Upper portion of ovary and style with the distal tips of the septa and the rim around them. x 146

1 Top view of young stigma consisting of the tips of the septa (Se) and a rim surrounding them. x 85

2 Mature flower. x ca. 4

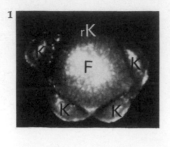
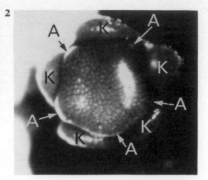
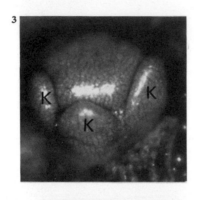
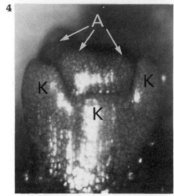
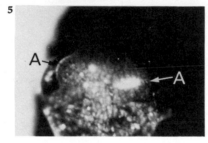
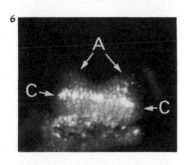
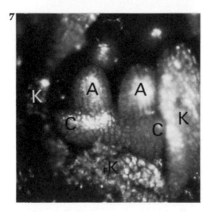

30 / PRIMULALES

PRIMULACEAE

Lysimachia nummularia L.
(creeping loosestrife)

Floral diagram
Note: The flower is subtended by a foliage leaf.

Floral formula: ✳ K(5) [C(5) A(5)] G(5) O∞
Note: The calyx and corolla tubes are so little developed that they are hardly noticeable in the mature flower. Externally the number of carpels cannot be determined.

Sequence of primordial inception
K1,2,3,4,5, A1-5, C1-5, G$_{girdling}$, O many in basipetal sequence

DESCRIPTION OF FLORAL ORGANOGENESIS

On the floral apex the sepal primordia are initiated first in a rapid spiral sequence (1, 2). In later developmental stages some growth occurs in the regions between the sepal primordia, thus producing an inconspicuous calyx tube.

Immediately after the inception of the fifth sepal primordium, the stamen primordia are initiated (2-5). All of them appear at about the same time; however, it is possible that they continue the genetic spiral of the sepal primordia with a plastochron approaching zero. During or immediately after stamen inception, the regions between the stamen primordia become slightly elevated over the floral apex. Then a petal primordium is initiated at the outer (abaxial) side of each stamen primordium (6, 7). Subsequently, growth extends from one petal primordium to another, thus giving rise to a corolla tube (7-10). Since further growth of the corolla tube is limited, it remains inconspicuous and is hardly noticeable in the mature flower (22). The stamen primordia develop into stamens with introrse anthers (11, 12, 22).

Immediately after petal inception, the pistil wall appears as a circular primordium (8, 9). With the upgrowth of this primordium and much radial expansion in its basal portion, the ovary and style become discernible (13-17, 21). In some floral buds, the rim of the gynoecial primordium seems to show five inconspicuous lobes (13). It is very doubtful whether these lobes are due

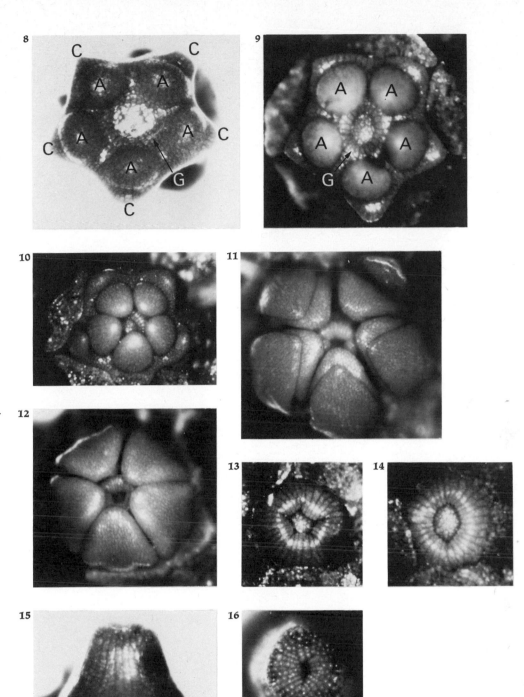

1 Top view of floral bud showing sepal inception (K). One sepal primordium was removed (rK). x 146
2 Top view of floral bud showing stamen inception (A). x 146
3 Same stage as that of figure 2, but seen from a different angle. x 146
4 Similar stage as that of figures 2 and 3, seen in side view. x 146
5 Similar stage as that of figures 2-4 after removal of sepal primordia, seen in side view. x 146
6 Side view of floral bud from which sepal primordia were removed to show petal inception (C) on two stamen primordia (A). x 146
7 Side view of more advanced floral bud; one sepal primordium was removed (rK). x 146
9 Top views of floral buds from which the young calyx was removed to show inception of the circular gynoecial primordium (G). x 146
12 Top views of floral buds showing stamen and petal development. The young calyx was removed. In figure 12 the petal primordia were also removed. Part of the gynoecial primordium is visible. x 85
16 Top views (13, 14, 16) and side view (15) of stages in early gynoecial development. Figures 13 and 14 show a slight variation of symmetry in the same developmental stage. x 146

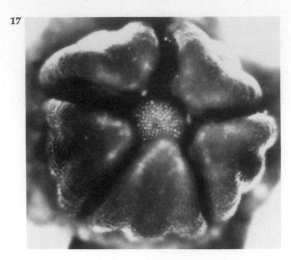

17

18

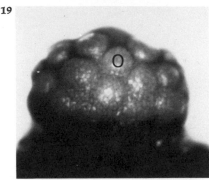

19

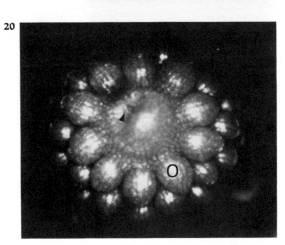

20

to more growth in five regions; they probably result from pressure of the developing stamens against the gynoecial primordium. Concomitantly with the upgrowth of the gynoecial primordium, the floral apex resumes growth and forms the free central placenta (18). Then ovule primordia are formed on the placenta in a basipetal (centrifugal) sequence (19). They alternate with each other in such a way that more or less distinct vertical and horizontal rows of ovules are formed (19, 20). Each ovule primordium develops two integuments. Before the formation of the integuments, the convex apex of the placenta (18, 19) resumes growth to form a projection which extends more or less into the base of the style (20). The style is rather long (21) and the original opening at its tip closes completely during the formation of the stigma (16, 17).

OTHER AUTHORS

I found no literature on the floral development of *Lysimachia nummularia*. However, some years ago, I reported on the floral development of *Lysimachia barystachys* Bunge (Sattler 1962). Apart from some differences in calyx aestivation, there is agreement in the floral development of

---

17 Top view of floral bud from which the young petals and sepals were removed to show young stamens and closure of the tip of the gynoecium. x 85

FIGURES 18-20. Stages of placental development. x 146

18 Side view of placenta before ovule inception.

19 Side view of placenta showing basipetal ovule inception. Only one ovule primordium is labelled (O).

20 Top view of more advanced placenta. One of the ovule primordia is labelled (O). The arrowhead indicates a slight damage of this preparation near the tip of the placenta.

21 Side view of nearly mature pistil. x ca. 18

22 Mature flower. x ca. 3

the two species of *Lysimachia*. Pfeffer (1872, quoted by Sattler 1962) comes to similar conclusions in a developmental study of *Lysimachia quadrifolia*.

Brulfert (1965) describes the floral development of *Anagallis arvensis*. According to her description and photographs, the gynoecial primordium is initiated shortly before or simultaneously with the petal primordia. Otherwise her results agree with my observations on *Lysimachia*.

For other literature before 1962 see Sattler (1962).

BIBLIOGRAPHY

Brulfert, J. 1965. Étude expérimentale du développement végétatif et floral chez *Anagallis* L., ssp. *phoenicea* Scop., formation de fleurs prolifères chez cette même espèce. Thèses. Université de Paris.

Sattler, R. 1962. Zur frühen Infloreszenz- und Blütenentwicklung der Primulales sensu lato mit besonderer Berücksichtigung der Stamen-Petalum-Entwicklung. Bot. Jahrb. *81*: 358-396.

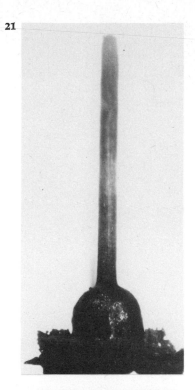

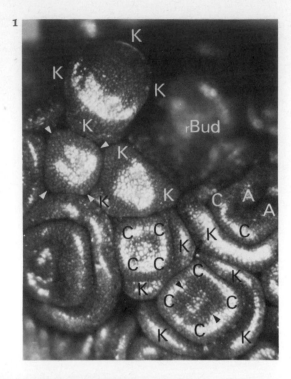

31 / OLEALES

OLEACEAE

Syringa vulgaris L. (common lilac)

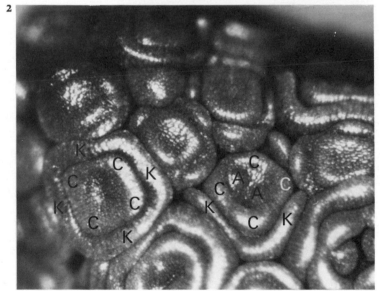

Floral diagram

Floral formula: $+ K(4) C(4) A_2 \underline{G}(2) O_4$

Sequence of primordial inception
K1-4, C1-4, A1-2, G1-2, O1-4

DESCRIPTION OF FLORAL ORGANOGENESIS

On the floral apex, the four sepal primordia are initiated at about the same time (1). It is possible that at least on some apices some of the sepal primordia appear in a very rapid sequence which approaches simultaneity. Since the floral apices and the floral buds are very crowded and in close contact with each other, slight deformations due to pressure may give the appearance of incipient primordia. Therefore, the exact origin of the sepal primordia is difficult to ascertain. One floral apex with only three sepal primordia was observed (1). Mature flowers with three sepals, three petals, two stamens, and two gynoecial appendages were also found. Immediately after inception of the sepal primordia, interprimordial growth occurs, thus initiating the calyx tube in a very early stage of development (1-5).

Very soon after sepal inception, the four petal primordia are initiated simultaneously (1, 2). Interprimordial growth occurs immediately after their inception or, in some buds, even during their inception (1, in the bud below the trimerous bud). In the latter case, the incipient corolla approaches a tetragonal girdling primordium. During further upgrowth, a long corolla tube with four distinct lobes is formed in all cases observed (12). In early and later stages of development the petal primordia grow faster than the sepal primordia.

Only two stamen primordia are formed simultaneously at the adaxial side of the very young corolla tube in positions alternate to the petal primordia (1 (see arrowheads on the bud at the bottom), 2). As the corolla tube grows upward, the developing stamens are carried up along with it. Each of them forms an introrse anther and a very short filament.

FIGURES 1-5. Portions of young inflorescence showing floral buds in various early stages of development. x 146

The bud on top shows sepal primordia (K) and a very faint indication of petal inception (not labelled). Below is one bud with three sepal primordia (K) only, and a young floral apex showing sepal inception (see arrowheads). The bud below the trimerous one shows a primordial corolla (C), and the bud at the bottom shows faintly stamen inception (see arrowheads). On the bud to the right stamen primordia (A) are well visible. One bud was removed (rBud). Note the partial overlap of buds.

In the two buds which are labelled the larger one to the left is still without stamen primordia whereas the slightly smaller one to the right has already distinct stamen primordia (A).

The labelled bud in figure 5 is a stage before gynoecial inception whereas the labelled bud in figure 4 shows the first indication of gynoecial primordia (G).

The labelled bud shows primordia of sepals (K), petals (C), and stamens (A). The gynoecial primordia are not visible in this undissected bud.

Top view of gynoecial primordia (G) of a floral bud from which the primordia of the stamens (rA) and the perianth were removed. x 146

Side view of more advanced gynoecial primordia (G). x 146

Top view of young dissected ovary showing the primordial septa (Se) with four ovule primordia (O) before integumental inception. x 146

Two slightly older ovule primordia. x 146

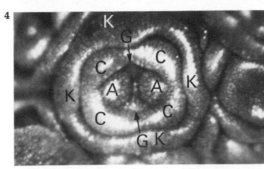

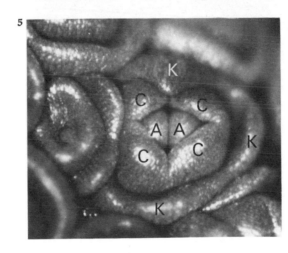

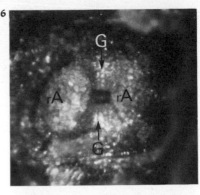

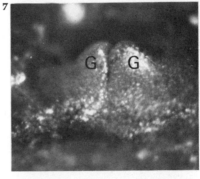

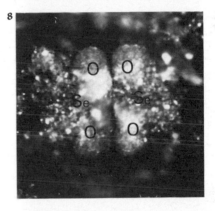

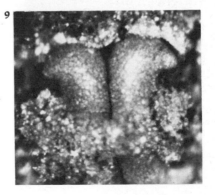

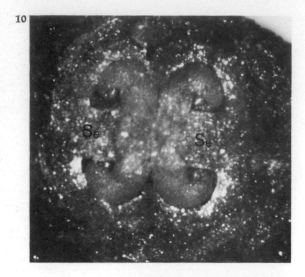

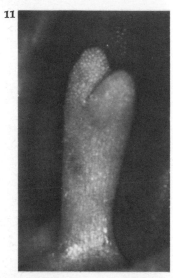

After stamen inception, two gynoecial primordia are formed in a plane at right angles to the plane of the two stamen primordia (4, 6, 7). The gynoecial primordia are located on the flanks of the concave floral apex at the base of the young corolla tube slightly below the insertion of the stamen primordia. As they grow upward, they become interconnected by interprimordial growth which results in a cylindrical portion. Further upgrowth of the cylindrical portion produces the ovary and style, whereas the gynoecial primordia form the two stigmas (11). Two septa grow out at opposing sides of the young ovary. Both form two ovule primordia on either side (8). In the center of the ovary the two septa come into contact and fuse postgenitally (8, 10). Each of the ovule primordia forms one integument and becomes anatropous (9, 10).

OTHER AUTHORS

Weber (1928) describes the organogenesis of normal and abnormal flowers of this species. At the time of sepal inception the floral bud is usually, though not always, elongated in the transversal plane. The median sepals arise slightly before the transversal ones, or vice versa. According to his figures on p. 620 all four sepals arise more or less at the same time, as reported here. In any case, the transversal sepal primordia grow faster than the median ones. The inception and development of the other appendages is in accordance with the description given here. Among the many abnormalities which he reports there are flowers in which the stamen may be replaced by one or two petals. Also the carpel may be substituted by a petal or an appendage intermediate between a stamen and carpel which may be more or less petaloid. In other abnormal flowers up to 10 additional whorls of petals may be present. In the case of, for example, one

additional whorl of petals, the plane of the carpels (styles) is changed by 90 degrees from the normal position. The additional whorls of petals may be alternating with each other, or two successive whorls may be opposite each other and in alternation with the preceding and succeeding ones. The number of petals (and sepals) per whorl may also vary. Torgad (1924; quoted by Weber) reports many other abnormalities in an extensive study. He worked, however, only with mature flowers.

Gracza's (1967) brief description of pistil development is in agreement with my results. He points out that in the early developmental stages the pistil arises as in an epigynous flower (e.g. the Umbelliferae), whereas in later stages the ovary becomes superior. For this reason, Gracza considers the ovary of this species in terms of development as a transitional type between the superior and inferior ovary.

King (1938) studied floral development of the olive which is very similar to that of lilac. He observes a perigynous tendency especially in the earlier stages of development. If a slight degree of perigyny occurs in lilac, it is quite inconspicuous.

BIBLIOGRAPHY

Gracza, P. 1967. Development of the pistil in *Syringa vulgaris*. Acta Agron. Acad. Sci. Hung. *49*: 439-442.
King, J.R. 1938. Morphological development of the fruit of the olive. Hilgardia *11*: 437-458.
Weber, G.F. Th. 1928. Vergleichend-morphologische Untersuchungen über die Oleaceen-Blüte. (Zschr. Wiss. Biol. Abt. E) Planta *6*: 591-658.

10 Top view of dissected ovary showing the two septa (Se) each one bearing two young ovules with one integument. x 85
11 Young style with the two primordial stigmas. x 86
12 Mature flower showing the corolla and the two stamens. x ca. 3.5

32/GENTIANALES (CONTORTAE)

ASCLEPIADACEAE

Asclepias syriaca L.
(common milkweed)

Floral diagram
Note: The pollinia are hatched inside the elongate stamen symbols. The corpuscula are drawn in black in the corners of the stigmahead. Translators, symbolized by simple lines, connect the corpuscula with the pollinia. The crests (horns) inside hoods of the corona are drawn as circles.

Floral formula:
$\ast$ K5 [C(5) A(5 with corona)] G{2} O∞
Note: The stamens and the stigmahead are postgenitally fused to form the gynostegium.

Sequence of primordial inception
K1,2,3,4,5, C1-5, A1-5, G, G1-2, O1-∞, Corona 1-5

DESCRIPTION OF FLORAL ORGANOGENESIS

On the floral apex the sepal primordia are initiated first in a spiral sequence (1). In later developmental stages little growth occurs between these primordia. The resulting calyx tube remains so small that it is not noticeable in the mature flower and is therefore not indicated in the floral formula and diagram. After the inception of the fifth sepal primordium, five petal primordia are initiated at about the same time (2). Subsequent growth between the petal primordia

leads to the formation of a corolla tube (5, 23). The free lobes of the corolla assume contort aestivation (18).

After the corolla, the five primordia of the androecium are initiated (2-4). Again, interprimordial growth occurs (4) producing an androecial tube (23). Some growth in a zone beneath the insertion of the stamen and petal primordia leads to the formation of a common base of androecium and corolla. However, this common base is not prominent in the mature flower. The anther region of the developing stamen primordium become slightly two-lobed. During the final stages of stamen development and the differentiation of two pollinia in each stamen, several complex growth phenomena occur. The tip of the stamen forms a scale-like projection which bends over the stigmahead (24). The sides of the anther regions form flaps which come in contact with each other (19, 22, 23) and are significant for the pollination mechanism (see Marcior 1965). At the outer base of the anther region a crescent-shaped hood primordium forms which at first is slightly more developed at its sides (19, 20). Concomitantly, a round horn primordium appears slightly above the crescent-shaped primordium (20). Both hood and horn primordia produce the conspicuous corona (22-24). Finally, a little projection develops on either side of the stamen, where the androecial tube connects the bases of adjacent stamens (21).

Long before the inception of the corona,

1 Top view of two floral apices (F) during inception of sepals (K). x 146
2 Floral bud from which all but one sepal primordium (K) were removed to show petal primordia (C). Stamen inception is faintly visible. x 146
3 Floral bud showing stamen inception (A). Young calyx was removed. x 146
4 Floral bud with stamen (A) and petal (C) primordia; young calyx was removed. x 146

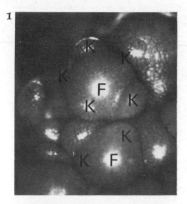

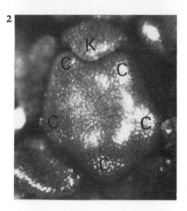

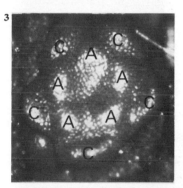

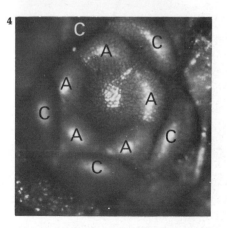

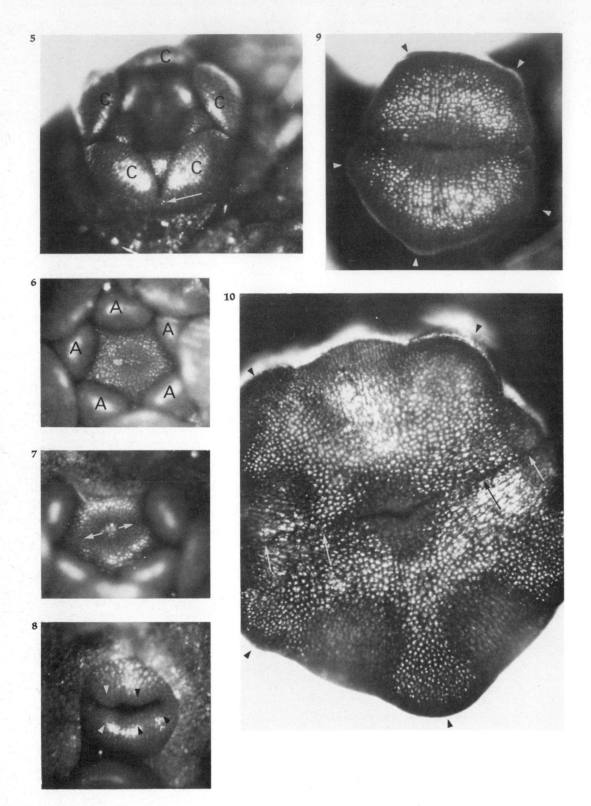

the gynoecium is initiated as a rim which completely surrounds the floral apex (6). The inner boundary of this gynoecial rim is circular, whereas its outer boundary is pentagonal due to additional growth in the gaps between the stamen primordia (6). It is difficult to decide whether the gynoecial rim is of equal height throughout, or whether it is slightly five-lobed due to slightly more growth in the regions which alternate with the stamen primordia. Immediately after its inception, the gynoecial rim stops growing in two regions. As a result, two gynoecial primordia emerge on the rim (7, 8). One of these gynoecial primordia seems to show two minute elevations, whereas the other one has three of them (see arrowheads in figure 8). These five inconspicuous elevations on top of the gynoecial primordia alternate with the stamen primordia. They are identical with and derived from the alterni-staminal regions on the gynoecial rim. After much upgrowth and development of the gynoecial primordia, the five elevations are still visible. On the differentiating stigmahead they even expand further into prominent lobes (9, 10). Five smaller lobes are formed between them (10). During the upgrowth of the gynoecial primordia, their margins grow inward until they meet in the center of the floral bud and fuse postgenitally (11, 17). Two pistils are formed by this process. They remain free from each other in the regions of the ovaries and styles. But in the region of the stigmahead they fuse postgenitally (9, 10) to such an extent that two entities are no longer discernible in the mature flower. The placentae develop on the ingrown margins of the gynoecial primordia. Many ovules are formed at about the same time with the exception of the uppermost ones which appear slightly later (15, 16). The ovules are arranged in vertical rows (16). Each of them becomes unitegmic and anatropous.

The differentiation of the corpuscula on

the lobes of the stigmahead, and the secretion of the translators, have not been studied.

OTHER AUTHORS

I found no recent literature on the floral development of *Asclepias*. Good's (1956) work on the Asclepiadaceae deals mainly with the structure of mature flowers. However, older literature on floral development by Corry (1881-87) and Dop (1903) is quoted. Frye (1902) describes briefly the floral organogenesis of *Asclepias* spp. Like most other authors, he deals mainly with the complicated later stages of development. Eichler (1875, p. 253) quotes a book by Schacht which includes a study on floral development of *Asclepias syriaca*. Unfortunately, this book was not accessible

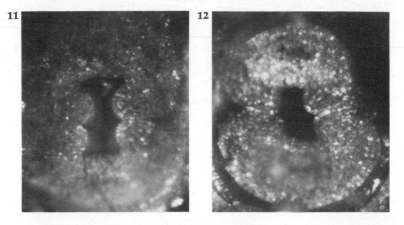

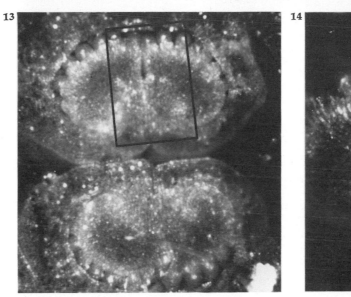

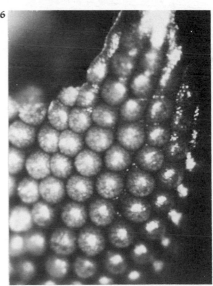

5 Side view of floral bud after removal of young calyx. Arrow indicates young corolla tube. x 146
6 Center of floral bud showing inception of gynoecium. x 146
7 Center of floral bud showing restriction of growth in two opposite regions of young gynoecium (see arrows). x 146
8 Two gynoecial primordia which were formed on the original gynoecial rim. Arrowheads point to five small elevations which are less clearly visible in figures 6 and 7. x 146
10 Top views of two stages in the development of the stigmahead. Arrows in figure 10 indicate postgenital fusion of the two gynoecial appendages. Arrowheads point to the corresponding regions of figure 8. x 146
12 Cross-sections through young gynoecium in its lower (11) and upper (12) portion. Note the incurving margins of the two gynoecial primordia. x 146
13 Cross-section through the gynoecium showing placentae and ovule primordia. x 85
14 Framed portion of figure 13 at higher magnification to show postgenital fusion of the two placentae of one gynoecial appendage ('carpel'). x 146
15 Lower portions of placentae with ovule primordia after removal of pistil walls. x 85
16 Placentae with ovule primordia. The upper portion of the left placenta was removed. x 146

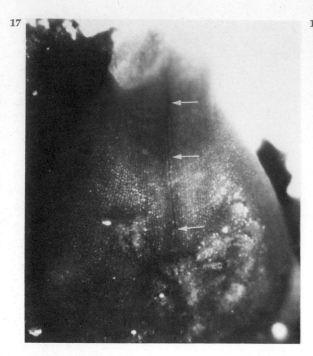

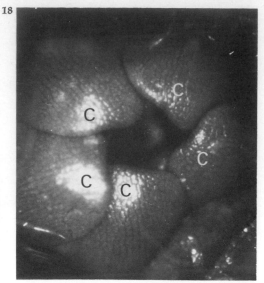

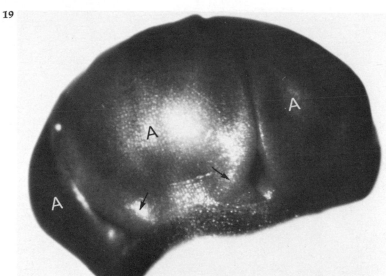

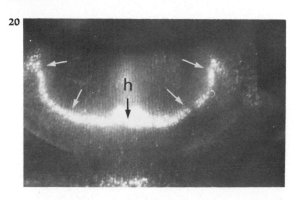

to me. Payer (1857) also describes the floral organogenesis of *Asclepias syriaca*. With the exception of the gynoecial rim, which he did not observe, his results agree rather well with my observations.

Recently, Nolan (1969) has described the developmental origin of the inflorescence in *Asclepias syriaca*. He did not study the development of flowers. Baum (1949) gives a brief account of gynoecial development with special reference to the origin of the placentae and ovules. Like Payer, she does not mention the original gynoecial rim, but otherwise her results agree with my observations.

Galil and Zeroni (1965) studied the production and distribution of nectar in the flower of a related species.

Puri and Shiam (1966) describe the floral anatomy.

BIBLIOGRAPHY

Baum, H. 1949. Die Stellung der Samenanlagen am Karpell bei *Asclepias syriaca*, *Cynanchum vincetoxicum* und *Erythraea centaureum*. Öster. Bot. Zeitschr. *95*: 251-256.

Eichler, A.W. 1875. Blüthendiagramme. Leipzig: W. Engelmann.

Frye, T.C. 1902. A morphological study of certain Asclepiadaceae. Bot. Gaz. *34*: 389-413.

Galil, J. and Zeroni, M. 1965. Nectar system of *Asclepias curassavica*. Bot. Gaz. *126*: 144-148.

Good, R. 1956. *Features of evolution in the flowering plants*. London: Longmans, Green & Co.

Marcior, L.W. 1965. Insect adaptation and behavior in *Asclepias* pollination. Bull. Torrey Bot. Club *92*: 114-126.

Nolan, J.R. 1969. Bifurcation of the stem apex in *Asclepias syriaca*. Amer. J. Bot. *56*: 603-609.

Payer, J.B. 1857. *Traité d'organogénie comparée de la fleur. Texte et Atlas*.

Paris: Librairie de Victor Masson.
Puri, V. and Shiam, R. 1966. Studies in floral anatomy. VIII. Vascular anatomy of the flower of certain species of the Asclepiadaceae with special reference to corona. Agra Univ. J. Research (Science) 15: 189–216.

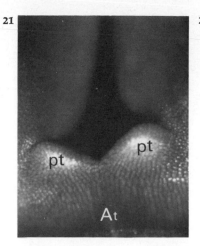

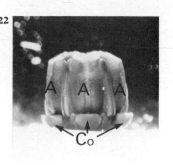

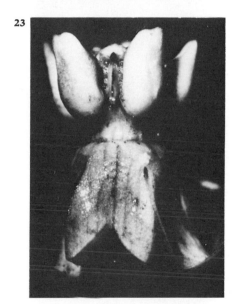

17 Inner side of one pistil ('carpel') showing post-genital fusion of incurved margins (arrows). For a cross-section of this developmental stage see figure 13. x 85

18 Top view of a young floral bud from which the young calyx was removed to show aestivation of corolla. This developmental stage corresponds roughly to that of figure 8. x 146

19 Side view of young stamens the central one of which shows inception of one corona hood (arrows). x 85

20 Base of a young stamen showing inception of hood (arrows) and horn (h). x 85

21 Androecial tube (At) in the region between two young stamens showing origin of two projections (pt). x 146

22 Side view of old floral bud from which calyx and corolla were removed to show the very late development of the hoods and horns of the corona (Co). x ca. 6

24 Side view (23) and top view (24) of mature flowers. x ca. 6

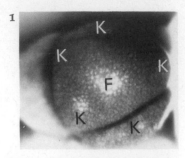

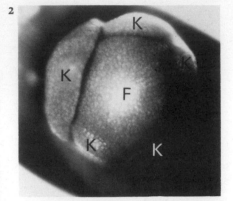

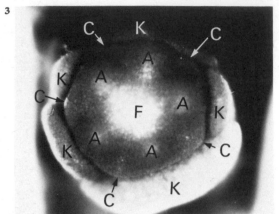

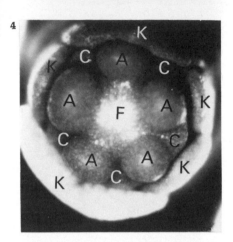

33/TUBIFLORAE

CONVOLVULACEAE

Calystegia sepium (L.) R. Br.
(= **Convolvulus sepium** L.)
(hedge bindweed)

Floral diagram
Note: The placenta is stippled; the disc is hatched.

Floral formula: ✳ K5 [C(5) A5] D(5) G(2) O4

Sequence of primordial inception
K1,2,3,4,5, C1-5, A1-5, G$_{girdling}$, G1-2, O1-4, D1-5

DESCRIPTION OF FLORAL ORGANOGENESIS

After the inception of two bracteoles, the five sepal primordia are initiated in a spiral sequence (1). This sequence may be clockwise (4) or counter-clockwise (1-3).

After the inception of the fifth sepal primordium, the floral apex assumes a pentagonal shape as a result of petal inception (2). The five stamen primordia are formed immediately after the petal primordia (3). They appear approximately at the same time; but a slight difference in the size of the primordia (3, 4) may indicate a very rapid sequential spiral inception. If this applies to petal inception too, the spiral sequence of the calyx would be continued by the corolla and androecium. In their

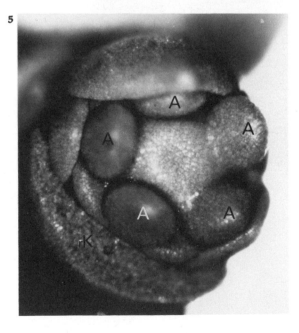

growth rate the petal primordia lag behind the primordia of the sepals and stamens (4, 5). Soon after the formation of the stamen primordia, growth between the initial petal primordia interconnects the petal primordia in the form of a ridge which completely encircles the androecium. This marks the beginning of the corolla tube formation. With further growth, the tubular portion of the corolla increases in height (18). Up to rather late developmental stages, the five corolla lobes are very conspicuous (17, 18). In the open flower, lobes are no longer prominent; the corolla is funnel-shaped (22). The characteristic aestivation of the corolla is shown in figure 19.

At a time when the stamen primordia have become about twice as high as the petal primordia, the gynoecium is initiated as a rim which completely surrounds the central floral apex in the form of a pentagon (5, 6). The five corners of the pentagon which are alternating with the five stamen primordia are more developed toward the outside, but they do not seem to be higher; the ridge seems to be of equal height throughout. However, shortly after its inception, the ridge shows more growth in two regions (7-9). As a result of this dif-

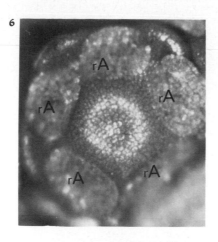

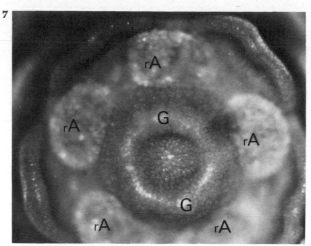

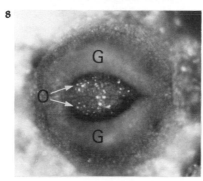

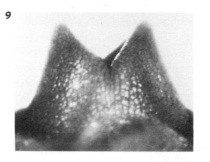

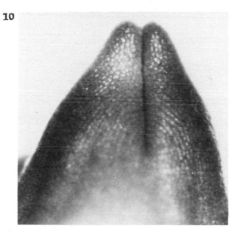

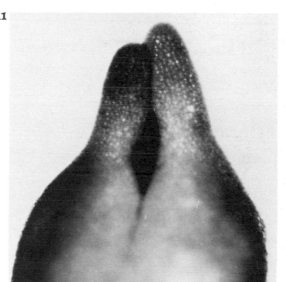

FIGURES 1-5. Top views of floral buds. x 146
1 Floral bud showing sepal inception (K).
2 Floral bud showing sepal inception and faint indication of petal inception.
4 Floral buds showing sepal (K), petal (C), and stamen primordia (A).
5 Floral bud from which the first sepal primordium was removed (rK) to show gynoecial inception.

FIGURES 6-11. Stages of gynoecial development. x 146
8 Top views of floral buds from which young sepals and stamens were removed (rA). Figure 6 shows the gynoecial ridge before the initiation of the two gynoecial primordia (G) which are shown in figures 7 and 8. In figure 8 arrows point to two incipient ovule primordia (O).
11 Side views of developing gynoecia showing the two gynoecial primordia and the cylindrical portion below them.

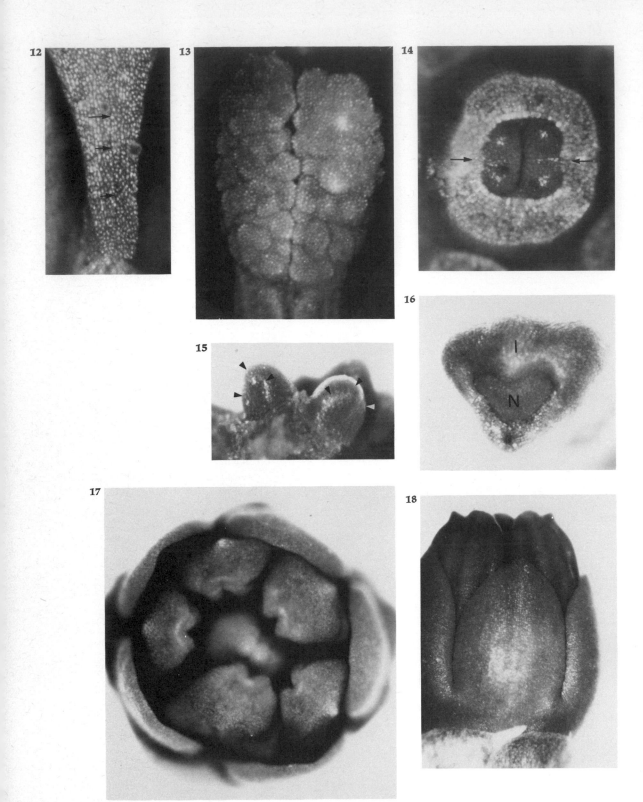

ferential growth, two gynoecial primordia arise (8, 9). Before the margins of these two primordia come in contact with each other and close the developing ovary, four ovule primordia become visible at the base of the ovary (8). As they develop, the central portion of the ovarial floor underneath them grows upward, thereby elevating them on a terminal placenta. Less growth occurs along a band at right angles to the gynoecial primordia; as a result, two pairs of ovules are separated by a depression in the placenta (14). Two double ridges (i.e. incomplete septa) develop from the margins of the developing gynoecial primordia (12). The latter give rise to the upper portion of the ovary, the style, and the two stigmas (10, 11, 13). Thus, the upper portion of the ovary and the style is a postgenital fusion product, whereas the lower portion of the ovary is formed by upgrowth of the pentagonal gynoecial primordium. The limit between these two parts of the ovary is at about the level where the double septa join the outer portion of the placenta through an incomplete single septum that is in continuity with the double septum above and the placenta below (14).

Each ovule primordium forms one integument while it attains its anatropous position. The integument is slightly three-lobed (15, 16).

After the inception of the ovules, the disc becomes visible as five primordia which soon form a five-lobed ridge surrounding the base of the ovary wall (20, 21). The five lobes of the disc alternate with the five stamens.

OTHER AUTHORS

I found no literature on the floral development of *Calystegia sepium*. Payer (1857) describes the floral organogenesis of a related species, *Convolvulus tricolor*. His results agree with my observations with a few minor exceptions. He reports that the gynoecium is initiated by two half-moon-shaped primordia which become connected by interprimordial growth. As his drawings show, the difference between this pattern and the one described above is only one of degree. Slight differences occur also in the position of ovule inception.

Frank (1876, p. 234) describes briefly the inception of perianth and androecium in *Convolvulus arvensis*. He notes a rapid spiral sequence in the initiation of sepals and stamens.

Hartl (1962, pp. 277-283) describes an 'apical septum' (Apikalseptum) in the upper portion of the gynoecium. He studied mainly mature or nearly mature flowers, and his description and illustrations refer to *Colonyction aculeatum*. However, he mentions the presence of such an apical septum also in *Calystegia sepium* L.

BIBLIOGRAPHY

Bhar, D.S. and Radforth, N.W. 1969. Vegetative and reproductive development of shoot apices of *Pharbitis nil* as influenced by photoperiodism. Can. J. Bot. 47: 1403-1406.

Brummitt, R.K. 1965. New combinations in North American *Calystegia*. Ann. Missouri Bot. Gard. 52: 214-216.

Frank, A.B. 1876. Ueber die Entwickelung einiger Blüthen, mit besonderer Berücksichtigung der Theorie der Interponirung. Jahrb. Wiss. Bot. 10: 204-243.

Hartl, D. 1962. Die morphologische Natur und die Verbreitung des Apicalseptums. Beitr. Biol. Pfl. 37: 241-330.

Payer, J.B. 1857. *Traité d'organogénie comparée de la fleur. Texte et Atlas.* Paris: Librairie de Victor Masson.

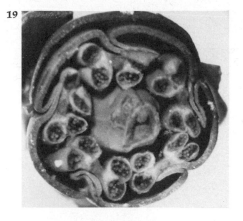

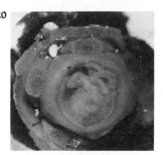

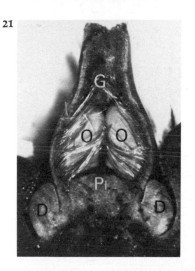

2 Inner surface of partial double septum showing postgenital fusion (see arrowheads) of the margins of adjacent gynoecial appendages. x 146

3 Side view of the two young appressed stigma. x 85

4 Top view of base of young ovary showing the two pairs of ovule primordia. This stage is slightly younger than that shown in figure 10. Arrows indicate the position of double ridges (septa) on the pistil portion which was removed. x 146

5 Side view of two ovule primordia. Arrowheads point to the three-lobed integument primordium. x 146

6 More advanced ovule primordium with three-lobed integument (I) and nucellus (N). x 146

7 Top view of floral bud from which the young calyx was removed. The gynoecium is at about the same developmental stage as that of figure 11. x 85

8 Side view of floral bud from which young calyx was removed to show young corolla and androecium. x 57

9 Cross-section through floral bud showing contortion of young corolla. x 13

10 Floral bud from which all organs were removed except pentagonal disc which surrounds the base of the pistil. x ca. 11

11 Longitudinally dissected floral bud showing disc (D), terminal placenta (Pl), two ovules (O) among hairs, and part of the pistil wall (G). x ca. 18

12 Mature flower with insects. x ca. 0.8

1

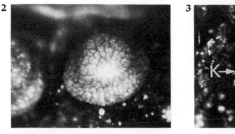

2

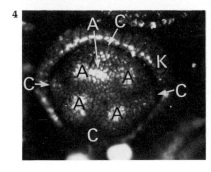

3

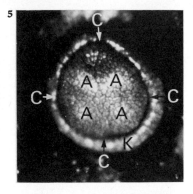

4, 5

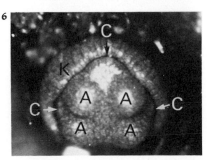

6

34 / TUBIFLORAE (SOLANALES)

VERBENACEAE

Lantana camara L.

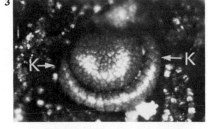

Floral diagram
Note: The asterisk indicates the position of the stamen primordium which usually does not develop into a structure noticeable in the mature flower.

Floral formula: $\cdot | \cdot$ K(2) [C(4) A4] $\underline{G}$(2) O2
Note: In the mature flower the calyx tube is slightly two-lobed; therefore it is symbolized as K(2), although it develops from a pentagonal rim (see below).

Sequence of primordial inception
K1-5, C1-4 A1-4,5, $G_{girdling}$, G1-2(3), O1-2

DESCRIPTION OF FLORAL ORGANOGENESIS

The calyx is initiated in the form of a pentagonal rim. That is, as the five sepal primordia arise, interprimordial growth occurs, producing a rim with a pentagonal outline (1, 2). In the adaxial and abaxial portion of this rim, less growth occurs than in its lateral portion. As a result of this, the original pentagonal rim becomes slightly two-lobed (3-8). On this two-lobed rim, the five corners of the original pentagon may remain faintly visible, two on one lobe of the rim, and three on the other lobe. However, due to slight irregularities in growth, the two lobes of the rim may become more or less undulated.

After the inception of the calyx, four stamen primordia are initiated (3-5). A fifth stamen primordium may appear slightly later in an adaxial position (4, 7). Growth of this stamen primordium is normally arrested in a very early stage of development. However, in some floral buds it develops into a complete stamen. In other floral buds, the fifth stamen primordium seems to be completely absent, even in its rudimentary form (5, 6). With further development, each stamen primordium forms a filament and a basifixed, introrse anther.

The petal primordia become visible immediately after the inception of the stamen primordia (4-6). Perhaps they are initiated simultaneously with the stamen primordia,

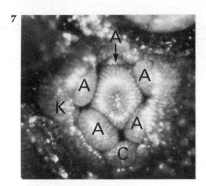

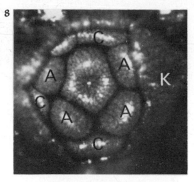

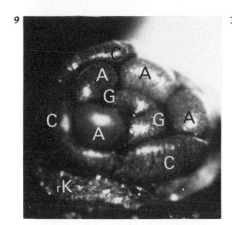

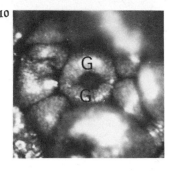

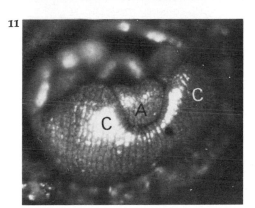

1 Top view of several floral apices showing the inception of sepal primordia (K) which are marked at one of the buds. F = floral apex. x 146

2 Side view of a floral apex with incipient pentagonal calyx. Note that the growth rate of the sepal primordia and the regions between them is almost identical, thus producing a pentagonal structure which is almost circular or tubular. x 146

3 Abaxial view of slightly older floral bud showing inception of androecium. The primordial calyx starts becoming two-lobed (arrows (K) indicate the two regions of enhanced growth). x 146

5 Top views of floral buds showing primordial calyx (K) and primordia of stamens (A) and petals (C). Only in figure 4 a fifth stamen primordium (→ A) is present. x 146

8 Top views of floral buds showing the tetragonal gynoecial rim in the center, stamen (A) and petal (C) primordia, and a slightly two-lobed primordial calyx (K). In figure 7 a fifth stamen primordium (→ A) is visible. x 146

9 Side view of a floral bud from which the young calyx was removed to show primordia of corolla (C), stamens (A), and inception of two lobes (= gynoecial primordia) on the gynoecial rim (= girdling gynoecial primordium). x 146

10 Top view of slightly two-lobed gynoecial rim. Each of the lobes is labelled as G. x 146

11 Side view of floral bud showing development of a corolla tube between the petal primordia (C); the young calyx was removed. x 146

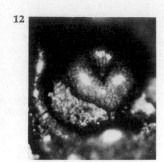

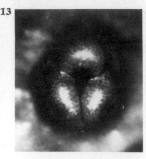

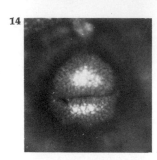

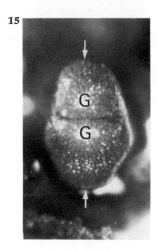

G
G

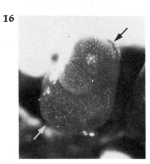

or even immediately before. However, no buds were observed which clearly showed an acropetal origin of the petal primordia. In floral buds without the fifth stamen primordium or with a fifth stamen primordium which does not develop, only four petal primordia are initiated; one of these is adaxial (opposite the rudimentary stamen, if it is present). In floral buds with five developing stamens, five petal primordia are present; two of the petal primordia are toward the adaxial side alternating with the adaxial stamen primordium, i.e., in a position comparable to that of the broad adaxial petal of the normal flower with only four petals. Growth extending between the petal primordia gives rise to a corolla tube (9, 11). Further growth in a zone at the base of the stamen and petal primordia leads to the stamens being inserted in the middle region of the corolla tube.

The gynoecium is initiated as a tetragonal rim (7, 8). The corners of this rim are alternate to the four developing stamen primordia. The adaxial corner of the rim which is opposite the non-developing stamen primordium is more prominent than the other three corners. During the upgrowth of the gynoecial rim, the adaxial and abaxial portions grow more than the lateral portions. Consequently, the rim becomes two-lobed (9, 10, 14). In some gynoecia three lobes develop instead of two (12, 13). During further development of the gynoecium, the lobes form the stigma, whereas the cylindrical portion below the lobes gives rise to the ovary and style. The placental primordia arise side by side at the abaxial portion of the developing ovary wall. Each develops into an anatropous unitegmic ovule (17, 18). A septumlike projection develops from the upper portion of the adaxial ovary wall. As it becomes two-lobed towards the top of the ovary wall it fuses completely with the placentae leaving a slit-like cavity between

the latter. (The placentae could also be referred to as septa.)

At the base of each developing stigmatic lobe a growth center originates (15, 16). Especially the abaxial one of these growth centers develops into a conspicuous mass of tissue which becomes bigger than the lobe on whose base it is formed (16). Also growth extends between the two growth centers, thus connecting the bulges by a ridge which subtends the original gynoecial lobes (10).

OTHER AUTHORS

I found no recent literature on the floral development of *Lantana*. Bocquillon (1861-62) studied the floral development of *Lantana camara* and other genera of the family. He reports acropetal succession of all whorls and only four primordia in the calyx, corolla and androecial whorl. Within the tetramerous whorls primordial inception proceeds from the abaxial toward the adaxial side. Payer (1857) describes the floral organogenesis of several species of Verbenaceae. He does not, however, deal with *Lantana*.

Junell (1934) studied the gynoecial morphology of *Lantana camara* and other related species and genera of the Lantaneae which are very similar to each other. He reproduces drawings and photographs of sections mainly through mature gynoecia. Both his text and illustrations agree with my observations.

Rao (1952) gives a short anatomical description of the mature flower of this species illustrated by five cross-sections at different levels. He concludes that one of the two carpels, the anterior one, is reduced and sterile. At its base, the ovary is unilocular; higher up it is divided by a septum which has a slit-like opening in its middle. According to my observations, this septum is the postgenital fusion product of the two placentae (which may also

be called septa) and the two-lobed septumlike projection from the adaxial side of
the ovary.

BIBLIOGRAPHY

Bocquillon, M.H. 1861-62. Revue du
groupe des Verbénacées. Adansonia 2:
81-165.
Junell, S. 1934. Zur Gynäzeumnorphologie
und Systematik der Verbenaceen und
Labiaten. Symb. Bot. Upsalienses 4:
1-219.
Payer, J.B. 1857. *Traité d'organogénie
comparée de la fleur. Texte et Atlas.*
Paris: Librairie de Victor Masson.
Rao, V.S. 1952. The floral anatomy of
some Verbenaceae with special reference
to the gynoecium. J. Ind. Bot. Soc. 31:
297-315.

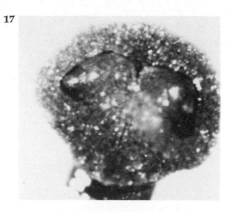

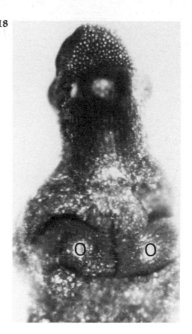

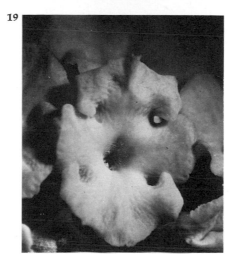

3 Young gynoecia with three instead of two lobes
(gynoecial primordia). x 146
6 Top views of developmental stages of two-lobed
gynoecia showing development of growth-centers
(see arrows) below the original gynoecial lobes
(G). Figures 14 and 15, x 146. Figure 16, x 85
7 Cross-section through young ovary showing separate insertion of the two ovule primordia (cf. floral
diagram). x 146
8 Side view of young pistil from which part of the
ovary wall was removed to show the two ovule
primordia (O). x 146
9 Top view of mature flower showing corolla only.
x ca. 4

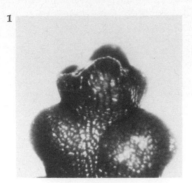

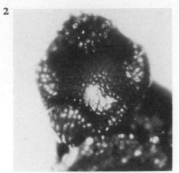

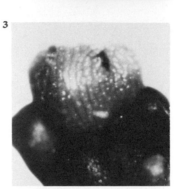

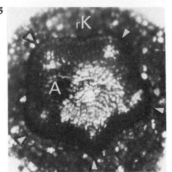

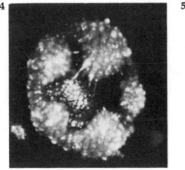

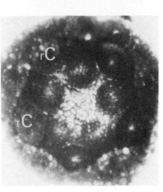

35/TUBIFLORAE

SOLANACEAE

Solanum dulcamara L.
(climbing nightshade)

Floral diagram

Floral formula: ✳ K(5) [C(5) A5] G̲(2) O∞
Note: The stamens are fused postgenitally in the anther region.

Sequence of primordial inception
K1,2,3,4,5, C1,2,3,4,5, A1,2,3,4,5, G1-2, O many basipetally (= centrifugally)
Note: The petal and stamen primordia may be initiated successively, but with a minute plastochron which approaches simultaneity.

DESCRIPTION OF FLORAL ORGANOGENESIS

The sepal primordia are initiated in a distinct spiral sequence (1-4). Both a clockwise (4) and counter-clockwise (2) direction of origin were observed. Before the inception of the petal primordia, a calyx tube arises by growth extending to the areas between the sepal primordia (3, 4).

The petal primordia are initiated at about the same time. However, a slight difference in their sizes seems to indicate that they appear successively with an extremely small plastochron. In contrast to the calyx tube, the corolla tube becomes distinct at a rather late stage of development (23).

After the petal primordia, the stamen primordia appear in succession with a minute plastochron (5, 6, 7). During their development, they form four-lobed anthers which fuse postgenitally, enclosing the pistil (24). Some growth occurs at the base of the corolla and stamens resulting in an inconspicuous common base of corolla and androecium (i.e. 'adnation' of stamens and corolla).

After stamen inception, two crescent-shaped gynoecial primordia arise simultaneously (8). Growth between these two primordia seems to occur immediately after their inception, giving rise to a cylindrical portion at the base of the two gynoecial primordia (8, 9, 10). This cylindrical portion grows very much and forms the ovary wall and style (15, 16). The two gynoecial primordia become appressed to each other and fuse postgenitally (12, 13, 14). This fusion is so perfect that it is no longer visible in older stages (14). Ridges form on the outer side of the pistil wall before the postgenital fusion of the gynoecial primordia is complete (15).

Two septa are initiated in a rather early stage of pistil development (10). These

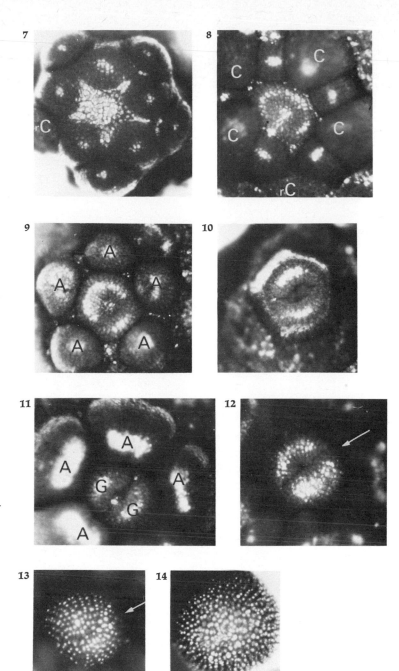

-4 Floral buds with sepal primordia. Figure 1 shows one terminal bud in side view and below two floral apices before or at calyx inception. Figure 3 shows the developing calyx tube in side view, figure 4 in top view. x 146

-8 Floral buds with the young calyx removed (rK) to disclose the inception of petals (figure 5, white arrowheads), stamens (6) and gynoecium (8). In figure 5 one stamen primordium (A) was formed. In figure 6 two petal primordia were removed. x 146

9 Slightly older stage than that of figure 8 whose developing calyx and corolla were removed. x 146

10 Developing pistil showing incipient septa. x 146

12 Parts of floral buds showing the tip of the developing pistil (and anthers in figure 11). The arrow indicates the plane of the septa. x 146

14 Top view of pistil tip showing the postgenital fusion of the two gynoecial primordia. In figure 14 the fusion is no longer detectable. x 146

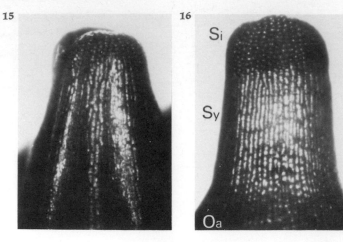

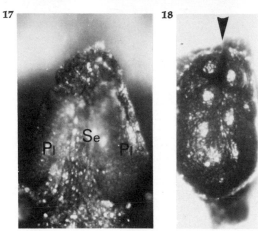

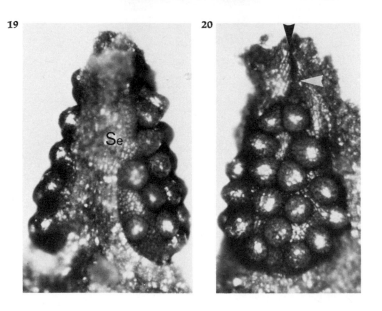

fuse postgenitally as they meet in the center of the pistil (21). Upgrowth of the floral apex (i.e. the floor of the gynoecium) in the region below the septa may occur, thus elevating the two septa on a common septum. This common septum, and the two septa which are continuous with it, enlarge toward the ovary wall and form the placentae (17). Ovule primordia are initiated in a rapid basipetal succession, i.e. the first ovules arise on the upper portion of the placentae (18, 19, 20). Each ovule primordium develops one integument.

Floral buds with an unusual numerical plan were observed. Some of these buds had three septa in the pistil (22), others had four or six petal primordia.

OTHER AUTHORS

To my knowledge the floral organogenesis of *Solanum dulcamara* has not been described. However, the floral development of other *Solanum* species and related genera has been reported (Efeikin 1958; Murray 1945 quoted older literature). Warner (1933, quoted by Murray 1945), studying *Lycopersicon esculentum*, reports a girdling calyx primordium on which the five sepal primordia are initiated. In contradiction to this, in *Solanum dulcama* the five sepal primordia are initiated before a primordial calyx tube is formed. Warner finds the same kind of origin for the corolla which in *Solanum dulcamara* is initiated also as five distinct primordia. Cooper (1927, quoted by Murray 1945), studying tomato, reports successive formation of the sepals, the petals, and the stamens. Young (1923, quoted by Murray 1945) mentions for the potato that 'the petals and stamens appear to arise simultaneously.' This might indicate that in the floral development of the genus *Solanum*, differences in the plastochron exist. With respect to the origin of the pistil, Young wrote: 'The pistil arises as a circular ring of tissue ...

Soon growth begins in the central part of the receptacle, forming the placenta.' Warner (1933) also reports a 'conical structure' in the center of the ovary before the inception of the septa. I did not notice a conspicuous central cone before septal initiation. Waterkeyn *et al.* (1965) describe the transition from the vegetative phase to the reproductive one in *Nicotiana* and illustrate it by photographs of whole buds and meristems.

BIBLIOGRAPHY

Efeikin, A.K. 1958. Development of the flower and raceme from the apical cone in *Solanum lycopersicum* L. Bot. Zhur. *43*: 1179-1183 (In Russian).

Murray, Mary A. 1945. Carpellary and placental structure in the Solanaceae. Bot. Gaz. *107*: 243-260.

Waterkeyn, L., Martens, P. and Nitsch, J.P. 1965. The induction of flowering in *Nicotiana*. 1. Morphological development of the apex. Amer. J. Bot. *52*: 264-270.

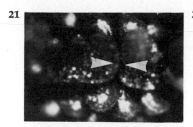

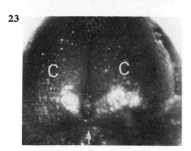

15 Side view of developing pistil showing the ridges in the lower portion. x 146

16 Side view of developing stigma (Si) and style (Sy) (most of the ovary (Oa) is omitted). x 146

17 Placentae (Pl) before ovule inception. The septum (Se) was detached from the ovary wall. x 146

18 Placentae of two postgenitally fused septa (see black arrowhead) showing the basipetal inception of ovule primordia. The placentae of figure 18 were photographed at right angles to the ones in figure 17. x 146

20 Different side views of placentate with ovule primordia. The arrowheads in figure 20 point to the fusion line of the septa. x 146

21 Cross-section through the septa with the placentae and ovule primordia. The white arrowheads point to the fusion of the septa in the center of the ovary. x 146

22 Unusual number of three septa (white arrowheads) instead of two. x 85

23 Side view of developing corolla. The white arrow points to the inconspicuous corolla tube. Calyx was removed. x 85

24 Mature flower showing the corolla, the androecium, and the stigma. x 2

36 / DIPSACALES

VALERIANACEAE

Valeriana officinalis L.
(common valerian)

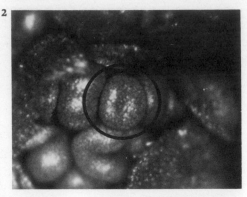

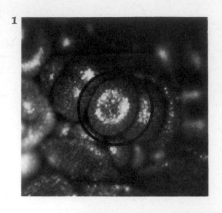

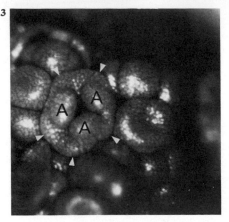

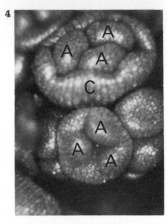

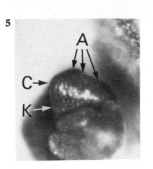

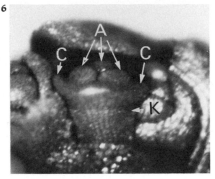

Floral diagram

Floral formula: K(ca.12) [C(5) A3] $\bar{G}$(3) O1
Note: The flower is asymmetrical due to the relative position of the stamens and the saccate spur at the base of the corolla tube.

Sequence of primordial inception
A1-3 C1-5, $K_{girdling}$, K1-ca.12, $G_{girdling}$, O1, G1-3

DESCRIPTION OF FLORAL ORGANOGENESIS

The flower develops as usual from a dome-shaped floral apex. The apex becomes concave due to increased growth on its periphery (1, 2). As this peripheral growth continues, it becomes strongly unequal: three regions grow upward, each one eventually developing into a stamen; and five regions outside the stamen primordia grow outward, developing into the corolla lobes (3, 4). Growth between and at the base of the corolla lobes produces the corolla tube (3, 4, 8, 19). This growth extends to a zone beneath the insertion of the three stamen primordia. Consequently, the stament primordia are gradually carried up on a cylindrical structure which may be called the 'corolla tube.' In the late stages of floral

organogenesis a sac-like evagination develops at the base of the corolla tube. In
addition, a septum is formed at the base of
the corolla tube (it is not indicated in the
floral diagram). This septum grows up to
the level where the stamens are inserted on
the corolla tube. Each stamen forms a filament and an introrse anther (19).

After the inception of the corolla and androecium, the calyx is initiated as a rim below the primordial corolla (5, 6). About 12
growth centers are formed on this rim (7).
Each of these curves adaxially as it develops
(17). Therefore, the calyx members are
hardly visible even in the mature flower.
They elongate drastically, however, during
the formation of the fruit. The number of
calyx primordia was found to vary from 11
to 13 in the 10 floral buds which were examined. Most of the buds had 12 calyx
members.

After the inception of the calyx (and
probably also after the inception of the

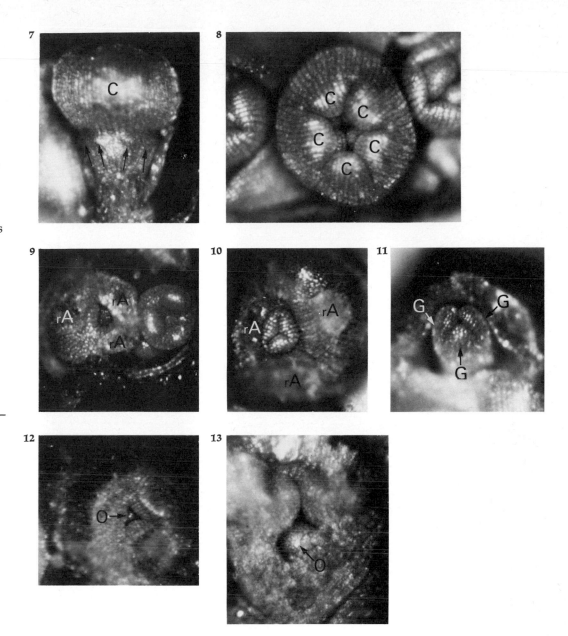

FIGURES 1-4. Top views of young partial inflorescences showing floral buds in different stages of
development. x 146

-2 Encircled are concave floral apices which do not
yet show distinctly the primordia of the androecium and corolla.

3 Floral bud with the primordia of stamens (A)
and petals (arrowheads).

4 One more advanced floral bud with primordial
corolla tube (C) and lobes, and three stamen primordia (A). Another bud like that is shown in
figure 3.

6 Side views of a floral bud with the primordia of
stamens (A), corolla (C), and calyx (K). x 146

7 Side view of floral bud showing young corolla (C)
and calyx rim after the inception of growth centers (arrows). x 146

8 Top view of floral bud showing young corolla (C)
and part of stamen primordia. x 146

1 Floral buds from which the corolla and stamens
were removed (rA) to show the development of
the triangular gynoecial rim and the three
gynoecial primordia (in figure 11). x 146

2 Top view of gynoecial rim. The ovule primordium
can be seen through the central opening. x 146

3 Young ovary from which part of the wall was
removed to show the ovule primordium (O). x 246

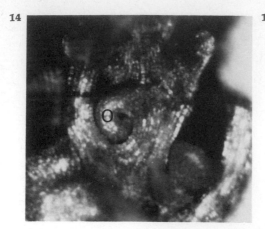

14

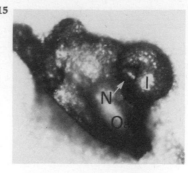

15

16

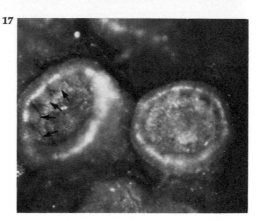

17

calyx members), a triangular gynoecial rim is formed on the flanks of the triangular cup-shaped floral apex. In descriptive terms one could also say that a girdling gynoecial primordium is formed on the inner side of the triangular corolla tube beneath the insertion of the stamen primordia (9). A triangular gynoecial tube develops from the gynoecial rim. Three primordia arise on this tube opposite the three stamen primordia (10, 11). These three primordia form three stigmatic lobes, whereas the gynoecial tube elongates into a style (19). The inferior ovary is formed by growth in a ring-zone below the insertion of the gynoecial tube. An ovule primordium is initiated near the base of the young ovary opposite one face (not a corner) of the triangular gynoecial primordium (12, 13). Obviously the growth which leads to the formation of the inferior ovary occurs mainly below the insertion of the ovule primordium. Consequently, the developing ovule becomes 'carried up' on the wall of the inferior ovary (14), and finally it is inserted near the top of the ovary (18). It forms one integument (15) and becomes anatropous (18).

During the development and displacement of the ovule, two septa are formed at the level where the ovule is borne. Consequently the ovary is three-chambered above the insertion of the ovule. The two chambers which are restricted to the ovarial portion above ovule insertion are very small (16, 18); they contain rudimentary ovules. The large chamber which extends over the full length of the ovary contains the functional ovule whose development was described above (18).

OTHER AUTHORS

I could find no recent literature on the floral development of this species or other species of the genus. However, Goebel (1898, p. 745) illustrated a few developmental stages by drawings of longitudinal sections showing the various growth zones. Asplund (1920) briefly describes gynoecial development of this species. He reproduces a cross-section through the upper portion of the ovary showing the fertile ovule primordium and two rudimentary ovules in each sterile locule. According to Asplund, the fertile ovule arises not on the ovarial wall, but on what he calls 'Zentralpartie' (central portion). This difference seems to be due to different terminology. What I called ovule primordium he terms 'Zentralpartie' whose upper portion later on becomes transformed into the ovule. Asplund describes in more detail gynoecial development in several species of *Valerianella*, some of which show three septa in the basal portion of the ovary. He quotes old literature on floral development of the Valerianaceae (e.g. Buchenau 1854). According to this literature, the stamens are initiated immediately after or simultaneously with the corolla.

Eichler (1875, p. 275) reports a variation in the number of calyx members from 12 to 20.

Payer (1857) studied the floral organogenesis of *Valeriana*, *Fedia*, and *Centranthus*. He illustrates, however, only that of the two latter ones.

BIBLIOGRAPHY

Asplund, E. 1920. Studien über die Entwicklungsgeschichte der Blüten einiger Valerianaceen. Kungl. Svenska Vet.-Akad. Handlingar 61: 1-66.
Eichler, A. W. 1875. *Blüthendiagramme*. Leipzig: W. Engelmann.
Goebel, K. 1898. *Organographie der Pflanzen*. 1. Teil. Jena: G. Fischer.
Payer, J.B. 1857. *Traité d'organogénie comparée de la fleur. Texte et Atlas*. Paris: Librairie de Victor Masson.

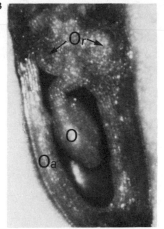

5 Longitudinally dissected young ovaries showing two stages in ovule development and displacement. I = integument, N = nucellus, Oa = ovary wall. x 146
7 Several buds from which the corolla with the stamens and style were removed to show the primordial calyx members on the calyx rim. Arrows point to some of the primordial calyx members. In figure 16 two of the floral buds were transsected in the upper part of the ovary to show the large chamber and the small chambers with the rudimentary ovules (Or). x 85
8 Longitudinally dissected ovary showing the large anatropous ovule (O) and the two small chambers with rudimentary ovules (Or) in the uppermost portion of the ovary. x 85
9 Mature flower. x 15

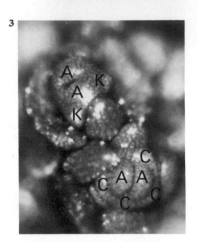

1

2

3

4

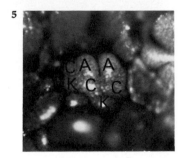

5

37 / CAMPANULALES

STYLIDIACEAE

Stylidium adnatum R. Br.

Floral diagram

Floral formula: ·|· K5 C(5) [A(2) $\overline{G}$(2)] O∞

Sequence of primordial inception
A1-2, K1,2,3,4,5, C1-5, O1-∞ G

DESCRIPTION OF FLORAL ORGANOGENESIS

The flower is formed from a dome-shaped floral apex. This apex gradually becomes concave. Thus, the androecial rim appears with two areas of enhanced growth, the stamen primordia (1-2, 6). At first, the stamen primordia grow much more than the rim. In later stages the rim elongates drastically, thus forming a long androecial tube (18).

Immediately after androecial inception (or perhaps even slightly overlapping with it), the five sepal primordia are initiated in a rapid spiral sequence (2-3). In their early stages of development they remain separate from each other. But in the later developmental stages a slight amount of growth occurs between them. This growth occurs mainly between the two sepals which are alternating with the reduced petal and between the other three sepals. Thus a slight partition in the transverse plane is produced which is, however, so inconspicuous that it is not indicated in the floral diagram and formula.

Soon after sepal inception, the petals are initiated at the same time (4-5). Growth between the petal primordia occurs immediately after their appearance (2-4). Thus, the corolla tube is initiated in the very early stages of development. In later stages it does not expand much, but it is clearly visible in the mature flower (18). The adaxial petal usually does not develop as the others do, but remains rudimentary (18). In some flowers, however, it had the same size, shape, and colour as the other four petals.

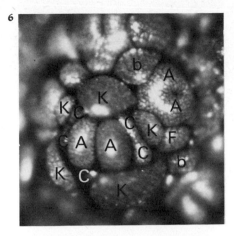

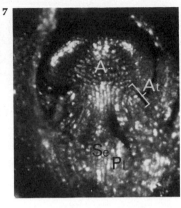

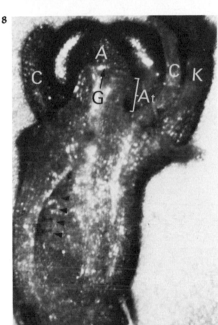

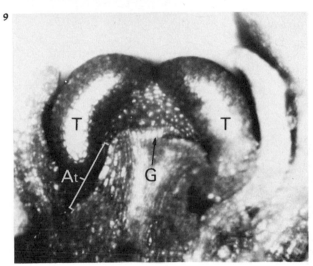

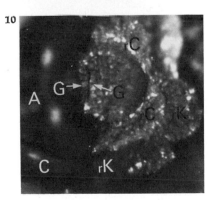

1 Floral apex which is becoming concave thus forming the two stamen primordia. Two bracteole primordia (b) are visible, one of them with an axillary floral apex (F). x 146

2 Inflorescence tip showing several floral buds in different developmental stages. The bud in the center shows the androecial primordia (A) and sepal inception (arrowheads). The more advanced bud below shows the five-lobed corolla (C) outside the two stamen primordia (A); a bud in a similar stage of development is seen in side view at the very tip of the inflorescence. x 146

3 Side view of a floral bud showing two of the sepal primordia (K) outside the androecial primordia (A). Top view of an older bud with incipient corolla (C). x 146

4 Side view of a floral bud showing inception of corolla (arrowheads). The bud may be slightly damaged on its right side. To the right is the young corolla (C) of another bud whose androecium was removed. x 146

5 Side view of floral bud showing petal inception (C). x 146

6 Top view of a floral bud with the primordia of sepals (K), petals (C), and stamens (A). To the right is a much younger bud with the two stamen primordia (A). x 146

7 Interior view of longitudinally dissected floral bud showing one stamen primordium, the young septum (Se) and placenta (Pl) before the inception of the stigma. At = young androecial tube. x 146

8 Same view as in figure 7 of an older bud. Arrowheads point to ovule primordia. The stigma (G) has been initiated. x 85

9 Upper portion of figure 8 at higher magnification. Note the two thecae (T) of the anther and half of the incipient stigma (G). x 85

10 Top view of an androecial tube from which one anther was removed to show inception of the stigma (G). x 85

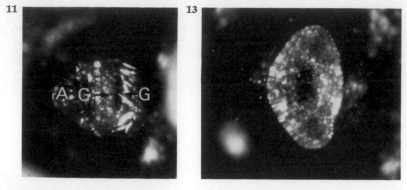

11

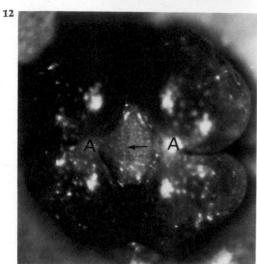

12

13

14

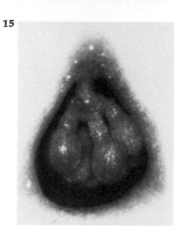

15

During the upgrowth of the androecial tube, growth occurs not only above the insertion of the primordial corolla, but also in a ring-zone below the primordial calyx. The cylindrical portion which results from the growth of the zone below the primordial calyx greatly enlarges and functions as ovary. A septum grows up from the base of this ovary and its walls at right angles to the median of the anthers (7). A placenta is formed only on one side of the septum. Many ovules are initiated on this placenta (8, 14-17). These ovules each develop an integument and become anatropous (14-17). No placenta and no ovules are formed in the second locule of the ovary (14). As the ovary enlarges, this second locule becomes more and more obscured until it is almost invisible in the mature ovary (15).

It should be noted that the inception of the septum, the placenta, and perhaps even the ovules occurs before the inception of any gynoecial primordium. The gynoecial primordium is initiated shortly after ovule inception (or perhaps simultaneously with it). It is formed on the inner sides of the two young bithecate extrorse anthers and on the upper rim of the androecial tube which connects the two anthers at their base (8-10). Since the inner sides of the two anthers are close together, and the androecial tube is rather oval in cross-section, the shape of the gynoecial primordium is also oval (13). As it grows upward it may become slightly two-lobed, i.e., the portions opposite the anthers may grow slightly faster than the regions alternating with them. But if this difference exists at all, it is quite inconspicuous. In any case, the two sides of the gynoecial rim which are opposite the anthers become appressed to each other and fuse postgenitally (11-12). Concomitantly, their surface becomes stigmatic (12, 18) and the fusion line is no longer recognizable. As a result, a seemingly solid stigma appears at the end of the androecial tube between the two anthers whose thecae are rather separate. It must be emphasized that only the stigma is formed by the gynoecial primordium. What functions as (and looks like) a style is the androecial tube which originates long before the stigma. Finally, the ovary itself is the lower portion of this tube or one may call it a cylindrical portion of the receptacle; it is definitely not formed by a gynoecial primordium or primordia.

OTHER AUTHORS

I could find no literature describing the complete floral development of *Stylidium adnatum*. However, Eckardt (1937) deals with the development of the ovary and stigma. His results are in agreement with my observations if one disregards the different terminology. According to his description the partition in the ovary originates from the placenta which eventually gives rise to a 'false septum.' With regard to the stigma, he mentions that no lobes are formed in the developing and mature stigma. Rosen (1935; quoted by Subramanyam 1951), who describes the embryology of *Stylidium adnatum*, reproduces a cross-section through two different ovaries. Only one of these cross-sections shows the rudimentary locule as my figure 14. The other cross-section may have been cut at the base of the ovary where the rudimentary locule is absent.

Eichler (1875, p. 301) quotes Schacht who claimed acropetal development and the presence of five stamen primordia, three of which degenerate during floral ontogeny. Barnéoud (also quoted by Eichler) observes, as I did, only two stamen primordia.

Subramanyam (1951) describes the floral organogenesis of *Stylidium graminifolium*, a species with many ovules in both locules. Many of his statements disagree with my observations and are not necessarily supported by his illustrations. Baillon (1876) studied the floral development of the same species. Most of his conclusions agree quite well with my results. The few discrepancies may be due to actual differences between *Stylidium graminifolium* and *S. adnatum*. Baillon also investigated *Stylidium adnatum*, but he mentions only that the early floral development is the same as in *S. graminifolium* and that differences occur only in later stages of development.

BIBLIOGRAPHY

Baillon, H. 1876. Traité du développement de la fleur et du fruit. XVI (Stylidées). Adansonia 12: 351-361.

Eckardt, Th. 1937. Untersuchungen über Morphologie, Entwicklungsgeschichte und systematische Bedeutung des pseudomonomeren Gynoeceums. Nova Act Leopold., N.F., 5: 1-112.

Eichler, A.W. 1875. *Blüthendiagramme*. Leipzig: W. Engelmann.

Subramanyam, K. 1951. A morphological study of *Stylidium graminifolium*. Lloydia 14: 65-81.

---

11 Same view as in figure 10 of a slightly more advanced stage. Hairs obscure partly the stigmatic lobe to the right. x 85

12 Top view of nearly mature androecium and stigma. The arrow points to the postgenital fusion line of the two stigmatic lobes which together with their cylindrical base form the stigma. Each of the two stamens (A) consists of two bisporangiate thecae. x 85

13 Cross-section through the stigma showing that its major portion is formed by upgrowth of an oval portion in continuation with the androecial tube. x 146

14 Cross-section through young ovary showing septum (Se) with the placenta (Pl) and ovule primordia (O). The rudimentary locule (Lo) is very small and becomes obscured later on (see figure 15). x 146

15 Cross-section through nearly mature ovary. The rudimentary locule is not clearly recognizable. x 85

16 Side view of flower bud from which a portion of the ovary wall was removed to show ovule primordia. x 85

17 Longitudinally dissected ovary showing placenta with ovules of an older stage than that of figure 16. x 85

18 Mature flower showing the androecial tube (At) with the anthers (A) and stigma (Si). Cr = rudimentary petal. Oa = inferior ovary. x ca. 10

16

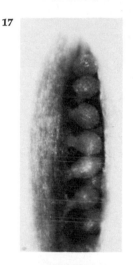

17

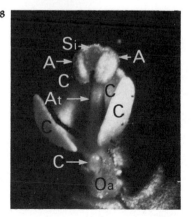

18

COMPOSITAE (ASTERACEAE)

Tragopogon pratensis L. (salsify)

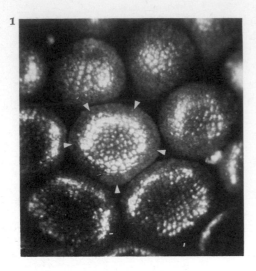

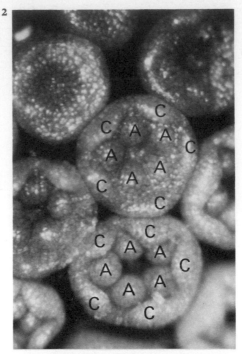

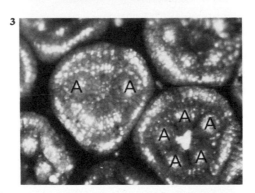

Floral diagram
Note: Because of less growth between the two corolla lobes toward the adaxial side than between the other corolla lobes, the flower is ligulate, i.e., zygomorphic.

Floral formula: $\cdot | \cdot\ K_{pappus}\ [C(5)\ A\langle 5\rangle]\ \bar{G}(2)\ O1$

Sequence of primordial inception
C1-5, A1-5, G1-2 Pappus in succession and overlapping in time of inception with O1.
Note: Probably the primordia of the corolla and androecium are initiated in very rapid succession.

DESCRIPTION OF FLORAL ORGANOGENESIS

On the floral apex, the corolla is initiated first. The periphery of the apex starts growing more than the center, resulting in the formation of a rim which surrounds the floral apex (1). As the rim develops, the five petal primordia appear due to differential growth. It is difficult to decide whether there are differences in growth rate on the rim from the beginning of corolla inception, or whether these differences appear after the inception of a circular corolla primordium (1). The petal primordia do not elongate very much; eventually they form the five teeth on the corolla (14). In contrast, the corolla rim grows a great deal with the exception of a minute adaxial portion (10, 11). In this way, it forms the complete ligulate corolla with the exception of the five teeth which originate from the five petal primordia.

The five stamen primordia are initiated soon after the petal primordia. They appear at about the same time; however, in at least some floral buds one can detect that they are formed in a very rapid sequence (3). The floral apex is concave when the stamen primordia are formed. Since the young corolla tube cannot be delimited from the concave floral apex, it is not meaningful to ask whether the stamen primordia are formed on the floral apex or on the adaxial base of the young corolla tube. As the stamen primordia develop, growth occurs in a ring-zone below the insertion of the stamen primordia and at the level of the young corolla. As a result, the stamen primordia are carried upward on the developing corolla tube (2, 4). Each stamen primordium develops into a filament and an introrse anther. The anthers of adjacent stamens fuse postgenitally, forming an anther tube around the style. The filaments of the stamens remain free.

The gynoecium arises as a two-lobed rim below the insertion of the stamen primordia (4, 5). Again, it is not meaningful to ask whether it is formed on the concave floral apex or at the base of the corolla tube. With further upgrowth, the two gynoecial primordia (i.e., the lobes of the gynoecial rim) develop into the two stylar branches with the stigmas. The rim on which the two gynoecial primordia are inserted elongates and forms the style. The ovarial cavity originates due to upgrowth of a ring-zone underlying the gynoecial rim and the developing corolla tube (13). The single ovule is initiated at the base of the ovary (13). It becomes unitegmic and anatropous.

At the time of gynoecial inception, or perhaps shortly thereafter, the first primordia of the pappus are initiated (7, 8). They alternate with the petal primordia, and their formation progresses from the adaxial side toward the abaxial side of the floral bud (7, 10). As the first five pappus primordia are initiated, growth occurs between them connecting them by a rim at their base (7, 8). On this rim more pappus primordia are formed (9-12). If one looks at the area between two of the original five pappus primordia, one may see first the emergence of two additional pappus primordia (10), then, outside these two, three other primordia may be initiated centrifugally (11). Growth occurs between the latter three primordia, and as a result, a second pappus ridge is formed on the outer side of the first one (11). This second ridge develops outside the first five alternipetalous primordia; hence, it encircles the whole floral bud. Besides this pattern, all sorts of variations occur after the first five alternipetalous pappus primordia have been

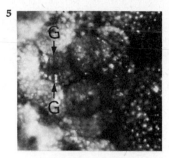

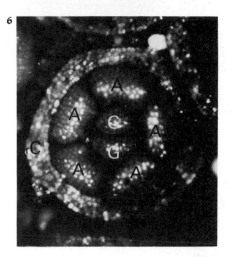

FIGURES 1-4. Portions of young heads showing floral apices and buds in various early stages of development. x 146

1 On one floral apex with the corolla rim the five sites of petal formation are indicated by arrowheads.

2 In two floral buds the stamen (A) and petal (C) primordia are labelled.

3 In the bud to the left the two labelled stamen primordia (A) are more developed than the three other ones, whereas in the bud to the right all stamen primordia (A) are at about the same developmental stage.

4 Some of the stamen primordia were removed (rA) to show gynoecial inception or a stage immediately preceding gynoecial inception.

5 Top view of dissected floral bud showing the two gynoecial primordia which are connected by interprimordial growth. x 146

6 Top view of floral bud from which the corolla was removed (rC) to show the primordia of the stamens (A) and stylar branches (G). x 146

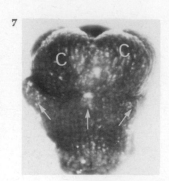

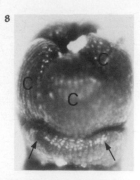

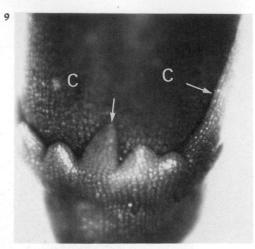

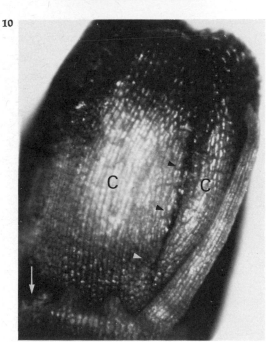

formed. The formation of additional pappus primordia is not necessarily as regular as described above (9, 12), especially towards the abaxial side of the floral bud. With further development, hairs are formed on the individual pappus primordia. Accordingly, the fruit has a plumose pappus.

OTHER AUTHORS

I found no recent literature on the floral development of *Tragopogon* spp. with the exception of Scheeffer-Pomplitz's (1957) study of the pappus. She notes, contrary to my observations, that the pappus is initiated before the corolla. In the last century, Warming (1876) described, besides many other Compositae, the floral ontogeny of *Tragopogon eriospermus*. His observations are very similar to mine. However, he reported that the first five pappus primordia are formed before the inception of the stamen primordia. The other pappus primordia do not appear in a definite order (p. 90); but Warming speculates about an average pattern of pappus inception (pp. 90-92). He reports also that the petal primordia are formed on a concave floral apex; this indicates that he considers what I termed corolla rim to be an evagination of the floral apex. Hence this latter discrepancy is interpretive (or terminological), not actual.

Jones' (1927) brief report on the floral development of lettuce (*Lactuca sativa* L.) agrees fairly well with my observations except that the pappus starts forming almost simultaneously with the stamens.

More recently, Lawalrée (1948) made a detailed study of floral development in several members of the Compositae. He did not investigate *Tragopogon*, but his observations on *Galinsoga, Bellis,* and *Chrysanthemum* agree in most respects with mine. Differences exist, of course, in pappus development. With regard to ovule inception, his drawings and photographs of sections show that the ovule primordium takes its

origin from the floral apex. Then, due to more growth on its adaxial side, the ovule becomes located towards the adaxial side of the ovary.

Older literature on floral development of other Compositae is quoted by Lawalrée and Warming.

BIBLIOGRAPHY

Jones, H.A. 1927. Pollination and life history studies of lettuce (*Lactuca sativa* L.). Hilgardia 2: 425-479

FIGURES 7-12. Side views of floral buds or parts of floral buds showing pappus inception and early development. x 146

7 Lateral view of floral bud. Arrows point to three of the first alternipetalous primordia of pappus members. The left arrow indicates the adaxial primordium.

8 Abaxial view of floral bud. Arrows point to the two primordia of pappus members which are alternating with the abaxial petal primordium (C).

9 Arrows point to alternipetalous primordia of pappus members, the one to the right being the adaxial one.

10 Two primordia of pappus members between the adaxial large pappus member and one alternipetalous smaller pappus member that has been removed (see arrow). Arrowheads point to the adaxial region of restricted growth in the corolla that will affect the ligulate corolla shape.

11 Similar view as in figure 10, but in an older bud after the centrifugal formation of three outer primordia of pappus members. Altogether the primordia of two alternipetalous and five antepetalous pappus members are visible. Arrowheads point to the same region as in figure 10.

12 Abaxial view of young pappus showing (in focus) the primordia of two alternipetalous and five antepetalous pappus members.

13 Longitudinally dissected floral bud showing the primordia of the basal ovule (O), one stylar branch (G), stamens (A), petals (C), and pappus members (arrows). x 85

14 Mature flower. Note the five teeth of the ligulate corolla which originated from the five petal primordia. x ca. 10

Lawalrée, A. 1948. Histogénèse floral et végétative chez quelques composées. Cellule 52: 215-294.

Scheeffer-Pomplitz, Marie-Edith. 1957. Morphologische Untersuchungen über den Pappus der Kompositen. Beitr. Biol. Pfl. *33*: 127-148.

Warming, E. 1876. Die Blüthe der Compositen. Hanstein's Bot. Abhandl. *3* (2. Heft): 1-167.

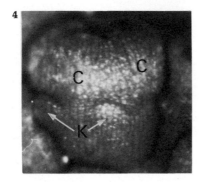

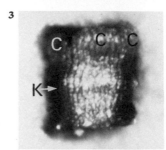

39/CAMPANULALES (ASTERALES)

COMPOSITAE (ASTERACEAE)

Tagetes patula L. nana
'Naughty Marietta'
(marigold)

Disc flower

Floral formula:
$* K_{(pappus)} [C(5) A\{5\}]$
$\bar{G}(2) O1$

Ray flower

Floral formula:
$K_{(pappus)} C(2-3) \bar{G}(2) O1$

Sequence of primordial inception
Disc flower: C1-5, K1,2-3,4-5 A1-5, G1-2, O1-2, D
Ray flower: C1-5, K1-2,3-4,5-6 A1-5, G1-2, O1-2, D
Note: In the ray flower particularly, at least some of
the petal primordia appear in an extremely rapid
sequence which approaches simultaneity. Pappus
inception overlaps with stamen formation.

DESCRIPTION OF FLORAL ORGANOGENESIS

Disc flower

The first primordia which are formed on
the floral apex are those of the petals. They
appear at about the same time and interprimordial growth occurs immediately after
their inception. Thus, the corolla tube is
initiated at a very early stage in development (1).

After inception of the corolla tube, the

adaxial primordium of the pappus is formed followed by the two lateral pappus primordia, and finally the two pappus primordia toward the abaxial side (2-4). Interprimordial growth between the five primordia occurs immediately after their inception (2-4). The adaxial primordium which is formed first grows much faster than the other ones (6-7). Therefore, it is longer than the others even in very late stages of development. Eventually, the margins of the pappus member develop hairs (8). Irregularities may occur in pappus development; however, no detailed study was made of the range of variation that occurs.

Overlapping with the inception of the last pappus primordia, the stamen primordia are initiated at about the same time shortly before the corolla encloses the floral apex completely. The stamen primordia are located at the base of the corolla tube. Due to growth in a ring-zone below their insertion, they are carried upward on the elongating corolla tube. Each stamen forms an introrse anther and a filament. The filaments remain free from each other, whereas the anthers fuse postgenitally.

After stamen inception, two gynoecial primordia are initiated on the deeply concave floral apex below (inside) the insertion of the stamen primordia (5). Interprimordial growth between the gynoecial primordia occurs immediately after or even during their inception. The cylindrical tube-like portion which results from interprimordial growth eventually forms a long style; whereas the two gynoecial primordia develop into the two stylar branches with the stigmatic surface. The inferior ovary is formed by upgrowth of a ring-zone below the insertion of the gynoecial primordia. The base of the young ovary grows upward to form two ovule primordia (9-12) one of which becomes a functional unitegmic anatropous ovule.

A ring-shaped disc is formed at the base of the style.

Ray flower
Five petal primordia are formed as in the disc flower (13), but only two or three of them develop further. The two primordia toward the adaxial side remain at the stage of most inconspicuous bulges (15, 16). The two lateral ones (which may be formed immediately before the other ones) grow upward quickly and form the two lobes of the mature corolla. The abaxial petal primordium develops into a small corolla lobe (22, 24) or becomes arrested at an early developmental stage so that it is no longer visible in the mature flower (23, 24). On the same head one may find ray flowers with and without an abaxial corolla lobe (24). Additional lobing may occur on some of the ray flowers (24). Thus, the same size of lobe may originate in quite different ways:

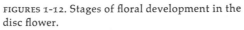

FIGURES 1-12. Stages of floral development in the disc flower.
1 Portion of a young head with its apex (R) and a number of centripetally developing floral buds and floral apices (F). On one floral bud the initiating petal primordia are marked with arrowheads. On two other floral buds the petal primordia are labelled (C). x 146
2 Floral bud showing petal primordia (note size differences!) and two primordia of the pappus (K). x 146
3 Side view of floral bud showing primordial corolla tube with petal primordia (C) and growth which will produce a small pappus tube. x 146
4 Side view of floral bud showing sequential origin of pappus primordia (K). x 146
5 Top view of two floral buds. From the bud to the right the young corolla was removed (rC) to show the stamen primordia. From the bud to the left, the stamen primordia were also removed to show the two gynoecial primordia (G). x 146
6 Side view of dissected floral bud showing the largest primordium of the pappus (K), a portion of the young corolla (C), and the primordia of two stamens (A) and the stylar branches (G). x 146
7 Side view of floral bud showing two primordia of the pappus (K). The one to the right is the adaxial primordium. x 146
8 Side view of an older floral bud showing a portion of the young pappus. x 85

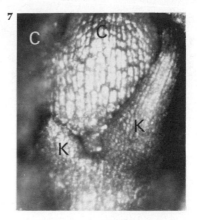

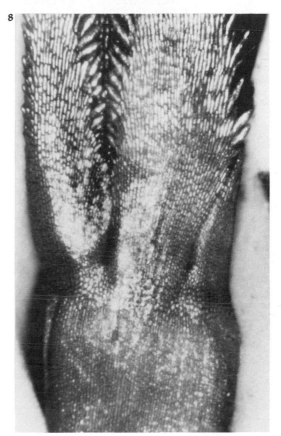

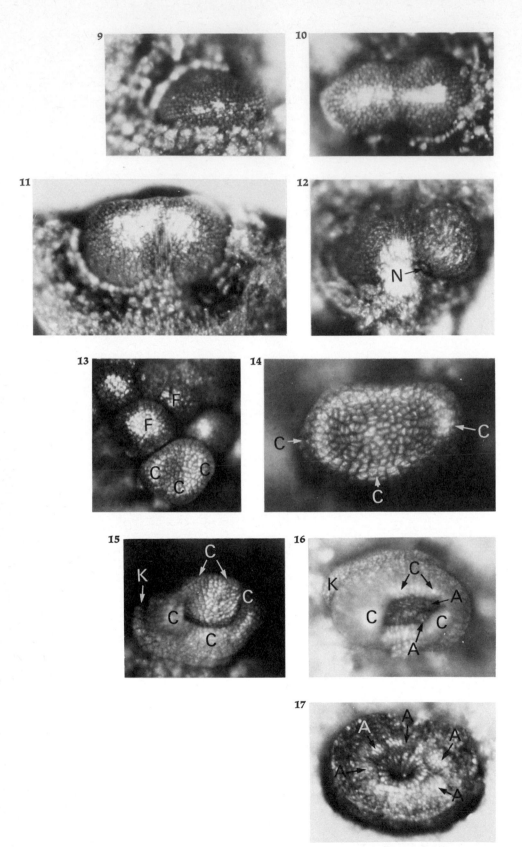

either from the abaxial petal primordium, or as a very late outgrowth at the margin of a lateral petal primordium. As in the disc flower, the corolla tube is initiated during or immediately after inception of the petal primordia. It may even overlap with petal inception, since in figure 14 the fifth petal primordium is not yet clearly visible whereas the corolla tube has been already initiated. The corolla tube develops much more toward the abaxial side than adaxially. As a result the ligulate form of the ray flower emerges (20, 24).

After the inception of the corolla tube, two primordia of the pappus are initiated opposite the two fast growing lateral petal primordia (15). Growth extends from these two primordia around the outer base of the young corolla tube. Concomitantly, two primordia originate toward the adaxial side (16) and two other primordia toward the abaxial side in positions alternate to the median petal primordium (18, 20, 21). Considerable variation in the upgrowth of the interprimordial regions and formation of new primordia on the original primordia produces a great variety in the structure of the mature pappus. There may also be variation in inception of the pappus primordia. No detailed study was made of the complexities of pappus development.

As in the disc flower, stamen inception overlaps with pappus inception (15, 16). The stamen primordia originally may be regularly distributed, but they soon become located in two groups opposite the two lateral petal primordia (17). This displacement is a result of lateral expansion of the whole floral bud which begins during corolla inception and becomes more prominent later on. The stamen primordia stop growing at a very early stage of development, and consequently, no trace of stamens is visible in the mature flower.

The development of the gynoecium and the disc (19, 20) is the same as in the disc flower.

I found no literature describing in detail the floral development of this species. Lawalrée (1948) and Warming (1876) quote many papers which deal with the Compositae in general or related genera. In some of these papers reference is made to *Tagetes*.

Sinha (1930) reports abnormalities in

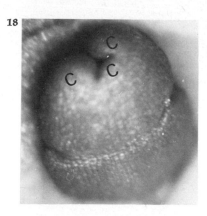

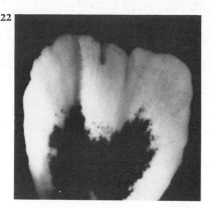

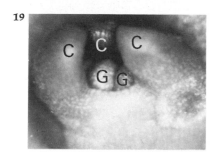

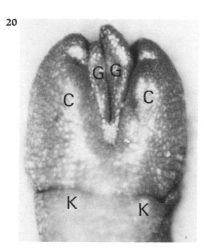

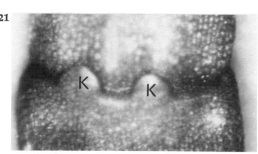

---

o Top views of placenta with the two ovule primordia. x 146

2 Side views of placenta with the two young ovules. In figure 12 the nucellus of one ovule is labelled (N). x 146

FIGURES 13-21. Stages of floral development in the ray flower.

3 At the bottom is a floral bud of a ray flower showing inception of the two lateral and the abaxial petal primordia. Above are two apices of disc flowers. x 146

4 Top view of a floral bud showing inception of petals and the corolla tube. One of the vestigial petal primordia toward the adaxial side is not yet clearly visible. The lateral petal primordia are marked with arrows, the abaxial one with an arrowhead. x 246

5 Floral buds showing the two vestigial petal primordia toward the adaxial side (marked with arrows), origin of the pappus (K), and the stamen primordia (A, in figure 16 only). x 146

7 Floral bud from which the young corolla was removed to show stamen primordia (A) and a first indication of gynoecial inception. x 146

8 Abaxial view of a floral bud showing the young corolla tube with the lateral and abaxial petal primordia. The primordial pappus rim is also visible. x 146

Adaxial views of floral buds showing the young corolla and the two gynoecial primordia (G). In figure 20 the two pappus primordia towards the adaxial side are visible (K). x 146

Abaxial portion of a floral bud showing the two pappus primordia toward the abaxial side (K). x 146

Corolla of ray flowers. In figure 22 the abaxial petal primordium has formed a little lobe in the middle, whereas in figure 23 no such lobe is present. In figure 23 the two stylar branches are visible. x ca. 3

Top view of a whole head showing six ray flowers and many disc flowers. x ca. 1.5

Tagetes erecta which showed an increase in the number of floral parts and various fasciations.

BIBLIOGRAPHY

Lawalrée, A. 1948. Histogénèse floral et végétative chez quelques composées. Cellule 52: 215-294.

Sinha, B.N. 1930. On a peculiar abnormality of capitula in *Tagetes erecta*. J. Ind. Bot. Soc. 9: 244-248.

Warming, E. 1876. Die Blüthe der Compositen. Hanstein's Bot. Abhandl. 3 (2. Heft): 1-167.

40 / HELOBIAE (ALISMATALES)

ALISMA(TA)CEAE

Alisma triviale Pursh.
(= **A. plantago-aquatica** var. **americanum** J.A. Schultes and Schult.)
(common water-plantain)

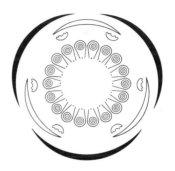

Floral diagram
Note: The margins of each pistil are only loosely appressed to each other.

Floral formula: ✳ K3 C3 A3×2 Gca.18 Oca.18 (as many as pistils)

Sequence of primordial inception
K1,2,3, CA1-3, C1-3, A1-6, G1,2,3, G$_{pistil}$ ca.18 in rapid succession, O as many as pistils in very rapid succession
Note: The pistil primordia (G$_{pistil}$) which are formed on six gynoecial primordia appear in rapid succession in a distinct pattern (see text). The same may be true for the primordia of all the other whorls including the CA whorl.

DESCRIPTION OF FLORAL ORGANOGENESIS

The first primordia which are formed on the floral apex are the sepal primordia. They appear in a very rapid succession. Soon after sepal inception, three bulges (i.e. CA primordia) appear in alternation with the sepal primordia. At the base of each of these bulges a petal primordium is initiated (1).

In contrast to the sepal primordia whose base is rather elongate in lateral direction, the petal primordia are not markedly dorsiventral; in fact, they have more the shape of stamen primordia (2, 3). However, in later stages of development, the petals become typically dorsiventral organs (14).

The six stamen primordia are formed in three pairs, each pair on a CA bulge above a petal primordium. Thus, the primordia of three petals together with those of the three stamen pairs alternate with the three sepal primordia (3, 5, 14). Each stamen primordium becomes first two-lobed, and then four-lobed as a result of anther formation (8, 9). The mature stamen consists of a filament and an extrorse anther. A slight amount of growth occurs between the stamen bases.

After the inception of the androecium, three gynoecial primordia are formed in alternation with the stamen pairs (3). In other words, the floral apex forms three bulges opposite the sepal primordia. Immediately afterwards, some additional growth may occur in the areas between the three gynoecial primordia, besides general expansion of the apex (4). The individual pistil primordia (G_{pistil}) are initiated first on the three gynoecial primordia (bulges) in the following sequence. Successively, one pistil primordium is formed in the center of each

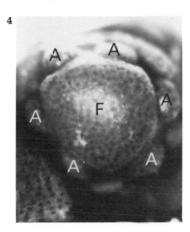
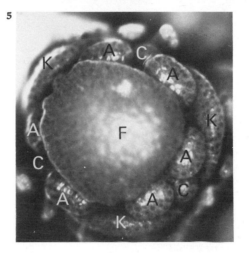
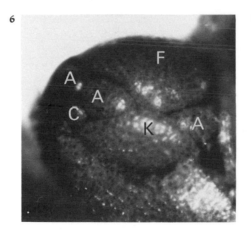
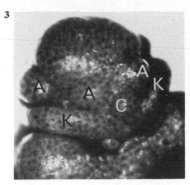
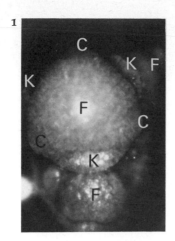
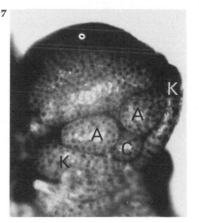
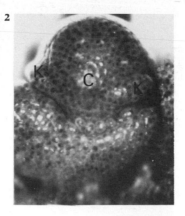

1 Top view of one very young floral bud and two floral apices (F) which have not yet formed appendages. The central floral bud shows the primordia of the three sepals (K) and the three petals (C). x 146

2-3 Side views of a floral bud before stamen inception (2) and a floral bud before the inception of individual pistil primordia (3). x 146

FIGURES 4-5. Top views of floral buds during gynoecial inception. x 146

4 Three antesepalous gynoecial primordia.

5 On the antesepalous gynoecial primordia pistil primordia are initiated in groups of three.

6-7 Side views of floral buds showing inception of pistil primordia. x 146

169 / *Alisma triviale*

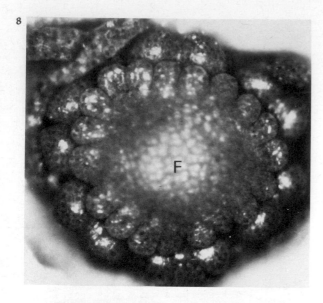

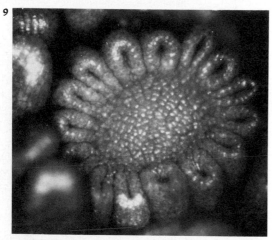

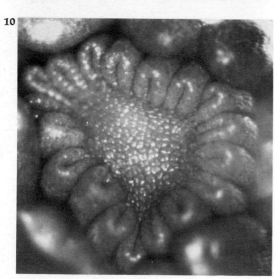

gynoecial bulge. Adjacent to each of these, two more pistil primordia are initiated, simultaneously one on either side. The sequence of the latter pistil primordia overlaps with that of the former (5). After nine pistil primordia have been formed in this fashion on the gynoecial bulges, additional pistil primordia are initiated in the three areas between the gynoecial bulges. First, two pistil primordia arise in each area, each of them near one of the lateral pistil primordia of two adjacent antesepalous groups. Then, often a pistil primordium is formed between the previous ones, i.e. in an exactly antepetalous position. If this latter one occurs in all three positions, the total number of pistil primordia is 18 (10). If it is absent in all three positions, the total number is 15 (9). If it is present in only one or two positions, the total number is 16 or 17 respectively. In some buds more than three pistil primordia are formed in an area between two adjacent gynoecial bulges. In these cases, the total number of pistil primordia is greater than 18 (see, e.g., figure 8 with 19 pistil primordia). The most common numbers of pistils in the material studied were 15-18.

The individual pistil develops as follows. At first the pistil primordium is a hump of tissue of a shape similar to that of a stamen or petal primordium (5-7). Then, at a very early stage of development, a region in the center of its adaxial side stops growing (8). As a result of this, a depression occurs which eventually represents the locule. The pistil wall develops from a horseshoe-shaped portion of the pistil primordium abaxial to the depression. As this region grows upward, its margins meet adaxially (9, 10, 13). Its upper portion develops into a style with a stigma, whereas the lower portion forms the ovary wall. The single ovule arises on the portion of the pistil primordium which is adaxial to the depression (11). As it develops two integuments, it becomes anatropous (12). The micropyle

points toward the abaxial base of the ovary wall.

OTHER AUTHORS

In the last century the floral development of *Alisma plantago* L. was described by Buchenau (1857) and Payer (1857). My results agree with most of the drawings of these authors, but there are some discrepancies with regard to their written statements. According to Payer the sepal primordia appear in sequence, whereas Buchenau seems to assume simultaneous inception of these primordia. These two statements need not be as different as they appear to be since the sequential appearance of these primordia is very fast and may indeed approach simultaneous inception. For this reason I cannot completely reject Buchenau's findings. A second discrepancy concerns the position of the stamen pairs. Payer states that the three stamen pairs are opposite the three sepals, whereas Buchenau claims that they are opposite the three petals, as confirmed here. With regard to gynoecial inception, Buchenau states that the floral apex becomes triangular before the 18-24 pistils are initiated. This statement on the triangularity of the apex is descriptively the same as saying that three gynoecial primordia are initiated. However, Buchenau overlooks the fact that the pistils are formed sequentially and in groups. Both he and Payer state that all pistils are formed simultaneously in one whorl.

More recently, Eckardt (1957) using microtechnical methods has described in detail ovule inception in *Alisma plantago-aquatica* L. and *Alisma gramineum* Gmel. Disregarding the different terminology, my observations agree with his results. For *Alisma plantago-aquatica* he reports a variation in pistil number from 9 to 27. Eber's (1933) description of epaltate carpels is contradicted by Eckardt's observations of a cross-zone as well as mine.

Daumann (1964) describes a disc at the base of the stamens which is formed by upgrowth in a zone underneath the stamen and sepal (not petal) bases. He also reports the presence of staminodia and transitional structures between stamens and staminodia in some flowers and variation in the number of stamens and staminodia (if the latter are present). Nectar is not produced by the disc, but by septal nectaries.

Singh (1966) studied the vascular anatomy of the mature flowers in the whole family of the Alismaceae.

For a more detailed study of the floral development see Singh and Sattler (1972).

Kaul (1967) describes the development and vascularization of the flowers of *Sagittaria latifolia* and *Lophotocarpus calycinus*.

BIBLIOGRAPHY

Buchenau, F. 1857. Über die Blütenentwickelung von *Alisma* und *Butomus*. Flora *15*: 241-256.

Daumann, E. 1964. Zur Morphologie der Blüte von *Alisma plantago-aquatica* L. Preslia (Praha) *36*: 226-239.

Eber, Erna. 1933. Karpellbau und Plazentationsverhältnisse in der Reihe der Helobiae. Flora *127*: 273-330.

Eckardt, T. 1957. Vergleichende Studie über die morphologischen Beziehungen zwischen Fruchtblatt, Samenanlage und Blütenachse bei einigen Angiospermen. Neue Hefte zur Morphologie. No. 3.

Kaul, R.B. 1967. Development and vasculature of the flowers of *Lophotocarpus calycinus* and *Sagittaria latifolia* (Alismaceae). Amer. J. Bot. *54*: 914-920.

Payer, J.B. 1857. *Traité d'organogénie comparée de la fleur. Texte et Atlas*. Paris: Librairie de Victor Masson.

Singh, V. 1966. Morphological and anatomical studies in Helobiae. VI. Vascular anatomy of the flower of Alismaceae. Proc. Nat. Acad. Sci. (India), Sect. B, *26*: 329-344.

Singh, V. and Sattler, R. 1972. Floral development of *Alisma triviale*. Can. J. Bot. *50*: 619-627.

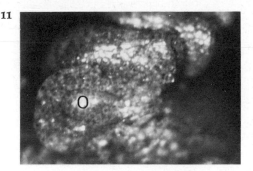

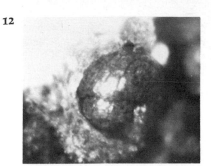

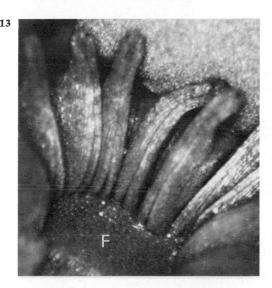

ɔ Top views of floral buds showing pistil development. In figure 8 the antesepalous pistil primordia show already the depression which will constitute the locule eventually, whereas some of the antepetalous pistil primordia are less advanced. x 146
1 Young pistils, one of them with part of the wall removed to show the ovule primordium. x 146
2 Ovule after inception of the two integuments. x 146
3 Residual floral apex and adaxial side of several young pistils. x 146
4 Mature flower. ca. x 5

41 / HELOBIAE (ALISMATALES)

BUTOMACEAE

Butomus umbellatus L.
(flowering rush)

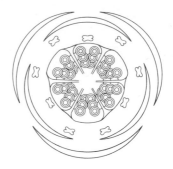

Floral diagram

Floral formula: $* \, P3+3 \, A3 \times 2+3 \, \underline{G}3+3 \, O\infty$
Note: The small amount of cylindrical upgrowth at the base of the gynoecium is not indicated in this formula (see floral diagram).

Sequence of primordial inception
$P_o1,2,3$, $P_i1,2,3$, A_o1-6, A_i1-3, $G1$-3,4-6, O many in very rapid acropetal succession
Note: The difference in primordial inception between the perianth and the inner whorls is only one of degree: in the inner whorls the plastochron between the primordia of one whorl—if it exists—is so small that it is difficult to ascertain, whereas in the perianth whorls it is still noticeable (at least in some floral buds).

DESCRIPTION OF FLORAL ORGANOGENESIS

The three outer tepal primordia appear in a rapid sequence on the floral apex. They are followed by the three inner tepal primordia which, in turn, are formed in rapid succession. Because of the minimal plastochron, it is difficult or impossible to determine clearly the direction of the sequence (1-3). The shape of inner and outer tepal primordia is similar. Differences occur mainly in later developmental stages (18).

After perianth inception, three pairs of stamen primordia are formed on the flanks above the inner tepal primordia (4-7). Primordia of adjacent staminal pairs are closer together than the primordia of the same pair (4). These are followed by three inner stamen primordia (5-7). There may be a very small plastochron between the inception of the three inner stamens as well as the three pairs of outer stamens, but if it exists, it is minimal (5, 6). The inner stamen primordia are inserted at a level considerably higher than that of the outer stamens. Correspondingly, the inner tepal primordia are at a higher level than the outer tepal primordia (1-3). Each stamen primordium forms a mature structure consisting of a filament and a latrorse basifixed anther (8-11, 18).

The six gynoecial primordia appear in a very rapid sequence in two whorls (9-11). The first three gynoecial primordia alternate with the three inner stamen primordia; the second set of three gynoecial primordia, at a level slightly inside the first three, are

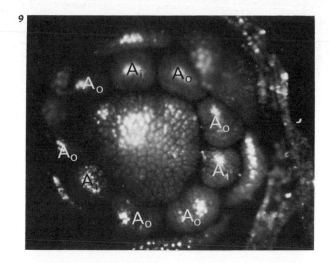

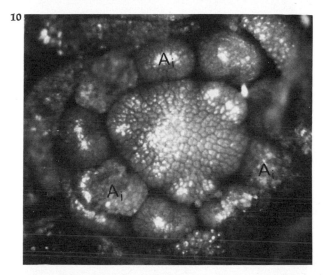

-3 Top views of floral buds showing tepal inception. x 146

4 Side view of a floral bud showing the inception of two stamen primordia belonging to adjacent pairs, opposite an outer tepal primordium. Inner tepal primordia are visible on each side of the outer one. x 146

-6 Top views of floral buds showing the primordia of the tepals and the outer six stamens. The primordia of the inner stamens opposite the inner tepals are just appearing (unlabelled). x 146

7 Side view of a floral bud in the same stage of development as those in figures 5 and 6. x 146

8 Side view of a floral bud after inception of the inner stamens (A₁). The outer three gynoecial primordia opposite the outer stamen primordia are just appearing (unlabelled). x 146

11 Top views of floral buds showing inception and early development of the six gynoecial primordia (pistil primordia). In figure 11 the dorsal portion of the pistil primordia is covered by the stamen primordia. x 146

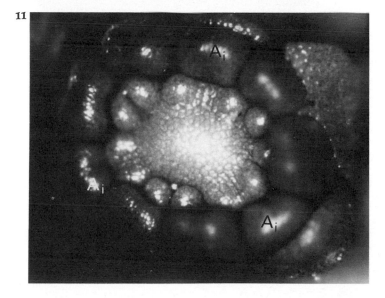

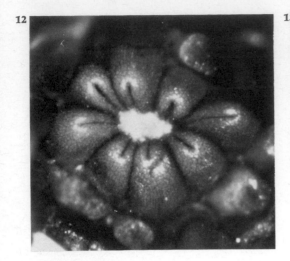

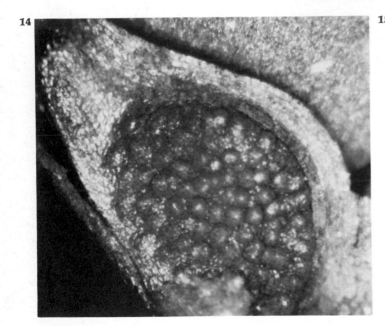

inside the three inner stamen primordia. Each gynoecial primordium develops into a pistil. It becomes horseshoe-shaped soon after its inception (10, 11). As it grows upward, the margins come in contact with each other, but they never fuse (12, 13). The upper portion of the young pistil forms an indistinct style with stigmatic margins; the lower portion develops into the ovary. There is some upgrowth in a ring zone below the abaxial side of the pistils. As a result, the six pistils are joined at their outer base for a short distance (17), otherwise they are free from each other. On the radial walls of each ovary, many ovule primordia are formed in very rapid acropetal succession (14-16). Each ovule forms two integuments and becomes anatropous (16).

During the later stages of development, some upgrowth occurs in a ring zone below the perianth, androecium, and the gynoecial wall. As a result of this, the flower tends to become epigynous. However, this epigyny is so inconspicuous that it is hardly noticeable and therefore it is not indicated in the floral formula and diagram.

OTHER AUTHORS

The floral organogenesis of this species is described by Buchenau (1857) and Payer (1857). My observations are almost in complete agreement with their results. Buchenau states, however, that the members of each whorl are initiated simultaneously. I noted a very rapid sequence of inception in all whorls. This sequence may be more or less pronounced. In some buds it is hardly noticeable. Therefore, differences in opinion may be due to variation in the material and the fact that we are dealing with borderline cases. Payer claimed sequential origin for the outer tepals, and simultaneous inception for the inner tepals. With regard to the stamens, he stated that the anthers are introrse. I termed the anthers latrorse; but since they tend to be somewhat intermediate between latrorse and introrse, the difference seems to be one of arbitrary classification. Schaeppi (1939) pictures them also as latrorse (figure 20, p. 413). With regard to the gynoecium, Buchenau claims central 'fusion' of the carpels. In contradiction to this, I could detect general upgrowth mainly at the periphery of the six carpels (see also Singh 1966).

Eckardt (1957) investigated the development of the gynoecium. My observations are in agreement with his results. He also emphasizes that the carpel margins are not fused. This explains why Johri and Bhatnagar (1957) find intracarpellary pollen grains in this species. Sprotte (1940) emphasizes that the carpels develop as epeltate structures.

Singh (1966) describes the vascular anatomy of the mature flower, and Roper (1952) reports on the embryology.

For more details on the floral development of this species see Singh and Sattler (in preparation).

Kaul (1967) describes the development and vascular anatomy of *Limnocharis flava*, a species which is a member of the Butomaceae but differs considerably from *Butomus umbellatus*.

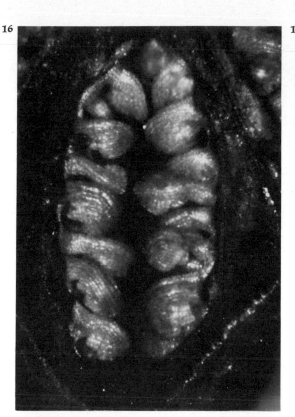

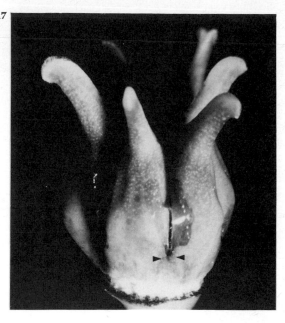

-13 Gynoecia of floral buds showing later stages of pistil development. x 85

14 Young pistil from which one half was removed to show ovule inception on the inner surface of the other half. x 146

15 Ovule primordia on a slightly older pistil than that of figure 14. x 146

16 Cross-section through a pistil of a mature flower showing the distribution of ovules along the radial sides of the pistil wall. There are no ovules at the dorsal portion and on the margins of the carpel. x 146

17 Gynoecium of mature flower consisting of six pistils (carpels) which have a common outer wall at their base (see arrowheads). x ca. 14

18 Mature flower. x ca. 5

BIBLIOGRAPHY

Buchenau, F. 1857. Über die Blütenentwickelung von *Alisma* und *Butomus*.
Flora *15*: 241-256.

Eckardt, T. 1957. Vergleichende Studie
über die morphologischen Beziehungen
zwischen Fruchtblatt, Samenanlage und
Blütenachse bei einigen Angiospermen.
Neue Hefte zur Morphologie. No. 3.

Johri, B.M. and Bhatnagar, S.P. 1957. Intracarpellary pollen grains in Angiosperms. Phytomorphology *7*: 292-296.

Kaul, R.B. 1967. Ontogeny and anatomy
of the flower of *Limnocharis flava* (Butomaceae). Amer. J. Bot. *54*: 1223-1230.

Payer, J.B. 1857. *Traité d'organogénie
comparée de la fleur. Texte et Atlas.*
Paris: Librairie de Victor Masson.

Roper, R.B. 1952. The embryo sac of *Butomus umbellatus* L. Phytomorphology *2*:
61-74.

Schaeppi, H. 1939. Vergleichend-morphologische Untersuchungen an den Staub-
blättern der Monocotyledonen. Nova
Acta Leopold. N.S. *6*: 289-447.

Singh, V. 1966. Morphological and anatomical studies in Helobiae. VII. Vascular
anatomy of the flower of *Butomus umbellatus* Linn. Proc. Ind. Acad. Sci. *43*:
313-320.

Singh, V. and Sattler, R. Floral development of *Butomus umbellatus*. In preparation.

Sprotte, K. 1940. Untersuchungen über
Wachstum und Nervatur der Fruchtblät-
ter. Bot. Arch. *40*: 463-506.

42 / LILIIFLORAE (LILIALES)

LILIACEAE

Allium neapolitanum Cyr.
(= Allium cowanii Lindl.)
(Naples garlic)

Floral diagram

Floral formula: $\ast$ P3+3, A3+3, $\underline{G}$(3) O6

Sequence of primordial inception
$P_oA_o1,2,3$, $P_o1 A_o1$, $P_o2 A_o2$, $P_o3 A_o3$, $P_iA_i1,2,3$
$P_i1 A_i1$, $P_i2 A_i2$, $P_i3 A_i3$, $G1-3$, $O1-6$
Note: P_oA_o = outer tepal-stamen primordium,
P_iA_i = inner tepal-stamen primordium.
The first inner common hump (P_iA_i) may be formed
before all of the P_oA_o primordia have given rise to
the outer tepal and stamen primordia.

DESCRIPTION OF FLORAL
ORGANOGENESIS

Three large primordia (P_oA_o) are initiated
first on the floral apex in very rapid succession, in clockwise or counterclockwise direction (1-3). Each of these primordia gives
rise to an outer tepal primordium which is
situated on the lower side of the common
hump, and an outer stamen primordium
situated on the upper side of the common
hump (3, 4, 5, 6). Following the outer large
humps, three inner common humps (P_iA_i)
form, varying between clockwise (4) and
counterclockwise (3) fashion. The first
inner common hump is initiated after at

least the oldest outer common hump has formed an outer tepal and outer stamen primordium respectively (3). Not all the floral buds attain the same size at corresponding developmental stages (1, 2). Occasionally a more advanced bud will be smaller in size than a less advanced bud (5, 6).

After all tepal and stamen primordia are well visible (7), three (rarely two) crescent-shaped gynoecial primordia appear, alternating with the inner stamen primordia (8, 10). Soon after their inception, the region between their margins starts growing (8, 9). However, this growth between the margins is very limited; it hardly contributes to the formation of the ovarial wall. The ovary and the style originate due to postgenital fusion of the gynoecial primordia (14). This fusion is so complete that it is no longer detectable in the mature style. Only the three stigmas remain separate. During the development of the pistil, the basal portions of the three gynoecial primordia bulge significantly toward the outside (16). As a result of this, the style appears to be overgrown and surrounded by the ovary.

Three placental primordia are formed at a very early stage of pistil development. They grow up with the floral apex which forms the center of the pistil (9, 10, 11). At

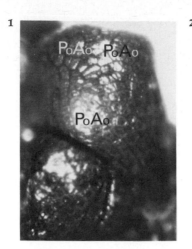

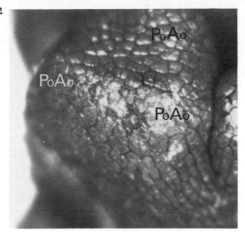

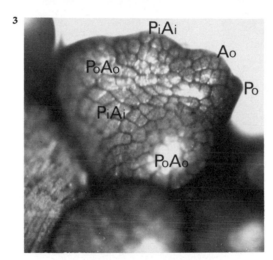

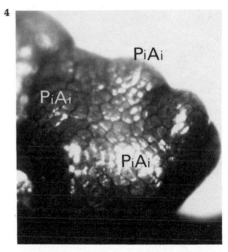

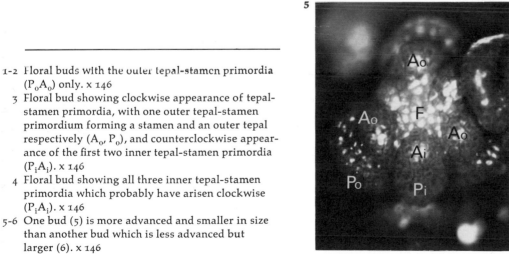

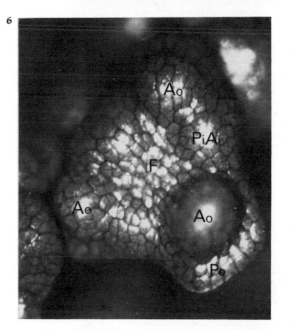

1-2 Floral buds with the outer tepal-stamen primordia (P_oA_o) only. x 146

3 Floral bud showing clockwise appearance of tepal-stamen primordia, with one outer tepal-stamen primordium forming a stamen and an outer tepal respectively (A_o, P_o), and counterclockwise appearance of the first two inner tepal-stamen primordia (P_iA_i). x 146

4 Floral bud showing all three inner tepal-stamen primordia which probably have arisen clockwise (P_iA_i). x 146

5-6 One bud (5) is more advanced and smaller in size than another bud which is less advanced but larger (6). x 146

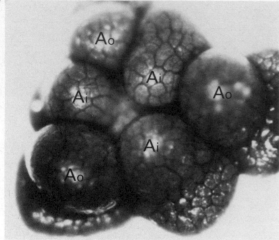

7

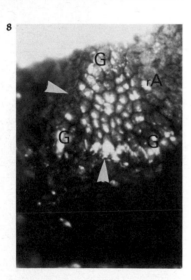

8

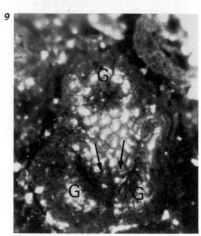

9

the same time there is continuity of growth between the incurved margins of the gynoecial primordia and the placental primordia (10, 11). Two ovule primordia develop on each placental primordium (12, 13). The ovules become bitegmic and hemianatropous with the micropyle pointing toward the base of the ovary wall (15).

OTHER AUTHORS

I found no literature on the floral development of *Allium neapolitanum*. However, Jones and Emsweller (1936) worked on the floral development of *Allium cepa*. Most of their results are in agreement with my observations. For example, they report that the stamen and tepal primordia are formed on common humps. But there are some differences between the two species. In *Allium cepa* the ovule position was indicated to be on the carpel margins. In *Allium neapolitanum* I found axile placentation. There may be differences with regard to the sequence in which the PA-primordia arise. However, the PA-primordia of one whorl may be formed in such rapid succession that it is often difficult to decide which primordium is the second or third one. Consequently, one cannot necessarily be certain whether a given sequence is clockwise or counterclockwise.

Schumann (1890), studying the early floral organogenesis of other *Allium* species, finds also common humps that give rise to stamen and tepal primordia. Schaeppi (1958) reports on the variation of mature flowers in a number of *Allium* species. Ustinova (1953) reports on floral abnormalities.

BIBLIOGRAPHY

Jones, H.A. and Emsweller, S.L. 1936. Development of the flower and macro-gametophyte of *Allium cepa*. Hilgardia *10*: 415-423.

Schaeppi, H. 1958. Untersuchungen über die Anzahl der Blütenblätter bei einigen Liliaceen. Bot. Jb. *78*: 119-128.

Schumann, K. 1890. *Neue Untersuchungen zum Blüthenanschluß*. Leipzig.

Ustinova, E.I. 1953. Anomalies in the construction of inflorescences and flowers of onions. Bot. Zhurnal *38*: 142–145 (In Russian).

---

7 Floral bud showing all inner and outer tepal and stamen primordia. x 146

FIGURES 8-11. Stages of gynoecial development. x 146

8 Three gynoecial primordia are visible (G) with little growth between them (see white arrowheads).

9 Three gynoecial primordia two of which are closer together in this bud. The placental primordia start forming (see black arrows).

10 Three placental primordia (arrows) at the base of the gynoecial primordia whose margins are close to each other (see white arrowhead).

11 Rare case of a gynoecium with only two primordia. Pl = placental primordium. Stamens were removed (rA).

FIGURES 12-15. Stages of ovule development.

12-13 Three placentae showing ovule inception. Most of the ovary wall (rOa) was removed. Figure 12, x 146. Figure 13, x 85

14 Side view of young pistil with portion of ovary wall removed. Arrowheads point to the fusion line between adjacent gynoecial appendages. x 146

15 Two bitegmic hemianatropous ovules of one locule. x 146

16 Mature flower. x 5

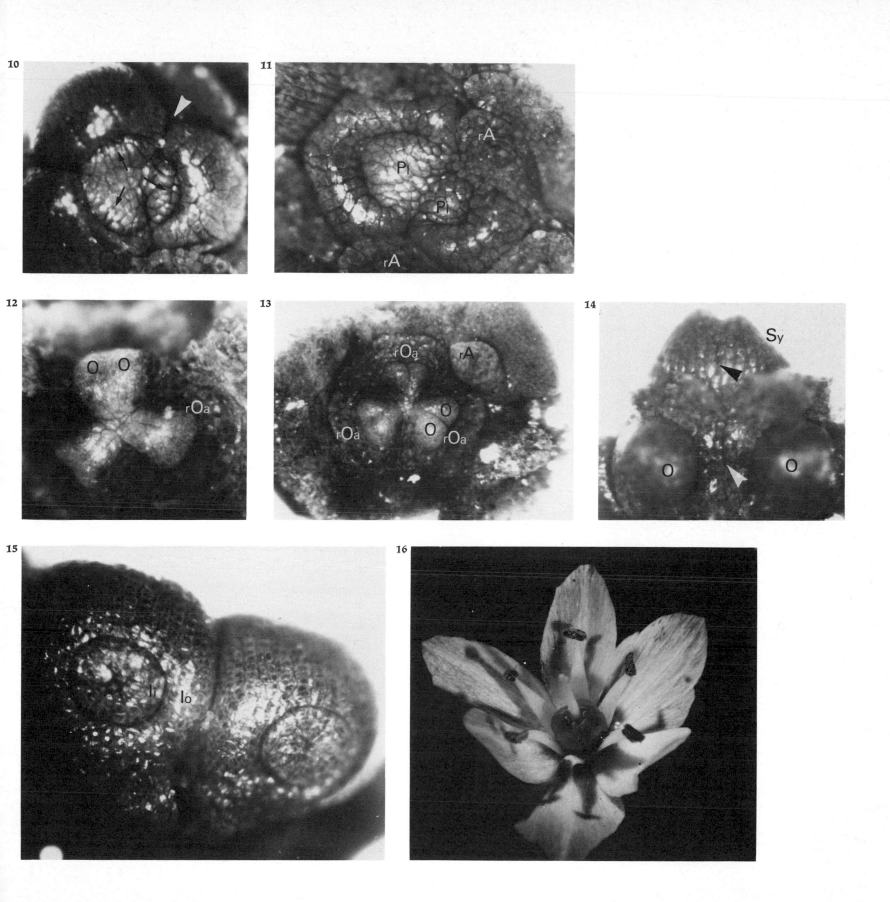

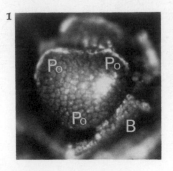

1

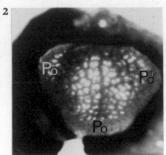

2

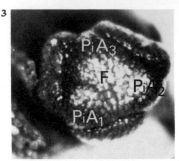

5

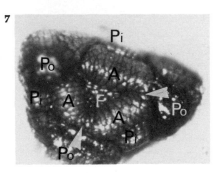

6

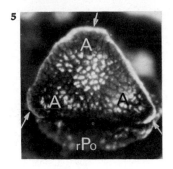

3

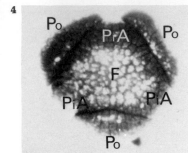

4

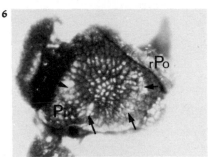

7

43 / LILIIFLORAE (LILIALES)

LILIACEAE

Ruscus hypoglossum L.

Floral diagram

Floral formula: $* P3+3 A(3) \underline{G}(3) O2$
Note: Instead of $\underline{G}(3)$ one could write $\underline{G}1$.

Sequence of primordial inception
$P_o1,2,3, P_iA1,2,3, P_i1 A1, P_i2 A2, P_i3 A3, G, O1, 2$
Note: $P_i A$ = tepal-stamen primordium. Primordia of one whorl arise in very rapid succession which may approach simultaneity.

DESCRIPTION OF FLORAL ORGANOGENESIS

The primordia of the three outer perianth members appear in very rapid succession; hence, it is difficult to determine their exact sequence (1). The abaxial outer tepal primordium is the last to appear (1, 2). Subsequently, three large humps, the inner tepal-stamen primordia, are initiated (3, 4). Each of these humps forms two distinct primordia: an inner tepal primordium, and, opposite this, a stamen primordium (3, 5, 6). It is possible, however, that the humps (P_iA) are the result of superposed tepal and stamen primordia with growth between them. Soon after the inception of the three stamen primordia, upgrowth occurs between them, leading to the formation of the stamen tube (6-9, 12-14, 23-24). Each stamen primordium becomes slightly two-lobed in lateral direction, at a very early stage of development (6-9). Subsequently, each of the stamen lobes forms another lobe on its outer side (12-14). By this lobing, each stamen develops an extrorse anther with four pollen sacs (14).

After the inception of the androecial tube, the gynoecium is initiated as a triangular primordium (8, 12). It is difficult to determine if slightly more growth occurs in the corners of the triangle (8-12). In some buds the outline of the gynoecial primordium is elliptical rather than triangular (11). This variation in shape of the gynoecial primordium appears to be correlated

---

-2 Floral buds with outer tepal primordia only. The abaxial tepal primordium is at the bottom. x 146

-4 Floral buds with the inner tepal-stamen humps, and the first appearance of inner tepal and stamen primordia at P_iA_i in figure 3. The inception of the common humps and their further development seems to be counterclockwise; the sequence of inception is indicated in figure 3 by numbers. x 146

FIGURES 5-9. Floral buds with the outer tepal primordia removed (rP_o) except two in figure 7. x 146

5 Each inner tepal-stamen hump formed a stamen primordium (A) on its inner side and a tepal primordium (arrow) on its outer side.

6 Two of the incipient stamen primordia are slightly two-lobed (the black arrows point to the lobes).

7 Upgrowth between the stamen primordia (see white arrowheads) starts the formation of the stamen tube.

8 Black arrowheads point to the corners of an incipient triangular gynoecial primordium.

9 Zygomorphic young androecium.

1 Gynoecial primordia: triangular (10) or elliptical (11). Arrows point to areas with perhaps slightly enhanced growth. x 146

FIGURES 12-14. Floral buds with the outer tepals removed. x 146

2 Arrowheads point to the corners of the triangular gynoecial primordium.

4 Small arrowheads point to the four lobes of a stamen primordium. The large arrowhead in figure 13 marks the growing region between adjacent stamen primordia.

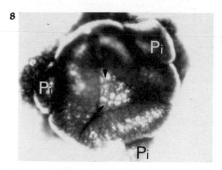

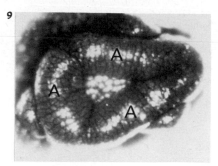

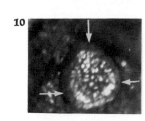

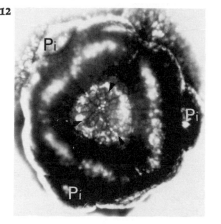

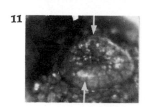

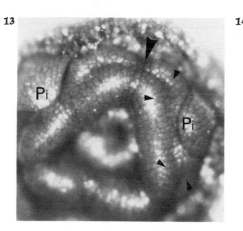

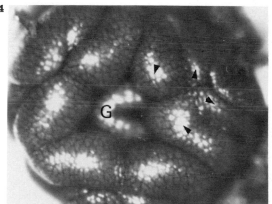

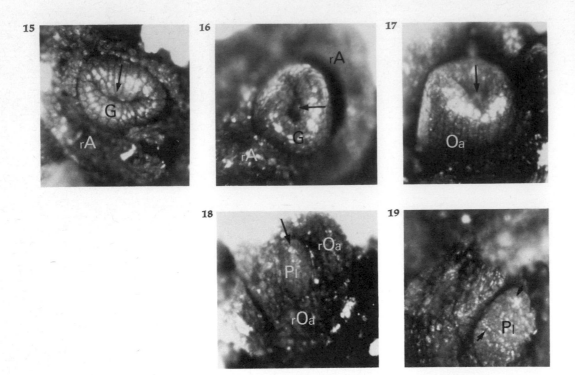

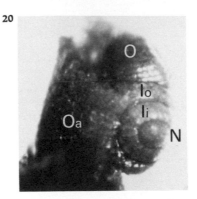

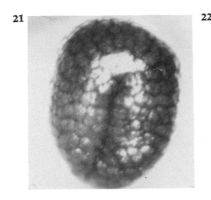

with variation in shape of the developing androecium (8, 9). General upgrowth of the gynoecial primordium results in a cylindrical structure (17) which will form the ovary and the style with either one or three stigmatic lobes. One stigmatic lobe develops when growth is restricted in the adaxial region only (21); three stigmatic lobes are formed in cases where growth is restricted in two additional areas toward the abaxial side (22). Transitional structures between the one-lobed and three-lobed conditions may occur (16).

The placental primordium appears on the adaxial side of the young pistil wall (15-17). The two ovules are initiated at opposite ends of the placenta (18, 19). They become bitegmic and anatropous (20).

OTHER AUTHORS

I found no literature on the floral development of *Ruscus hypoglossum* L. Savchenko and Dmitrieva (1962) describe the flower and the fruit of *Ruscus hypophyllum* L. They find that this species is semidioecious rather than dioecious: 'However in the process of further development, beginning from the stage of formation of the ovules and pollen, an inhibition in the development is revealed, in one case of the pistil, in the other of the stamens, which conditions the formation either of functionally male flowers, or of functionally female flowers. The frequent occurrence of examples bearing both functionally male and functionally female flowers, suggests that *Ruscus hypophyllum* L. is rather a semidioecious plant than a dioecious.' The same phenomenon is noted by Vigodsky (1936; see Savchenko and Dmitrieva 1962) in *Ruscus aculeatus* L. In *Ruscus hypoglossum* L., I found a well-developed gynoecium, but the anthers were reduced and dehisced before the flowers opened. I saw no pollen grains.

Savchenko and Dmitrieva continue in their paper by saying: 'The gynoecium of *Ruscus hypophyllum* L. consists of three carpels forming the superior syncarpous ovary. In the majority, the latter is usually unilocular, but sometimes the development of a partition is observed, as a result of which a bilocular or trilocular ovary is formed ... It possesses a short style, a large-headed stigma with a scarcely perceptible separation into three fringes.' In *Ruscus hypoglossum*, I never found septa, and the stigma consists of one or three lobes. Otherwise there is agreement between Savchenko and Dmitrieva's results on *Ruscus hypophyllum* and my observations on *Ruscus hypoglossum*.

Levacher (1968) describes the stamen and perianth inception in *Paris quadrifolia* L. He finds that the stamen and the opposite petal (perianth member) are initiated almost at the same time (figures 2, 3). Immediately after their initiation there is upgrowth between these two organs (figure 4). The result is a common stamen-petal complex (figure 6). It is possible that the same mode of origin occurs in *Ruscus hypoglossum*. No final answer can be given at the present time.

BIBLIOGRAPHY

Levacher, P.M. 1968. Sur deux particu-larités ontogéniques des étamines par rapport aux feuilles, sépales et pétales chez le *Paris quadrifolia* L. C.R. Acad. Sc. Paris *267* : 418-420.

Savchenko, M.I. and Dmitrieva, A.A. 1962. O Biologii razvitia *Iglitzi podlistnoi* (*Ruscus hypophyllum* L.) i nekotorikh osobenostei stroenia u tzvetka i ploda. (The biological development of *Ruscus hypophyllum* L. and certain particularities of the structure of its flower and fruit.) Trudy botnicsescovo instituta imeni V.L. Komarova Academii Nauk S.S.S.R., *5* : 45-57.

23

24

17 Stages of pistil development. The arrow points to the placental primordium. x 146

FIGURES 18-20. Dissected young ovaries showing stages of placenta and ovule development. x 146

18 Placenta with one ovule primordium (see black arrow).

19 Placenta with two ovule primordia (see black arrows).

20 Portion of ovary wall with ovule primordia, one of them showing the two integuments.

22 Tips of young pistils, one-lobed (21) or three-lobed (see arrows in figure 22); in both cases there was less growth on the adaxial side. x 146

24 Mature flowers. Figure 23 is a side view of the stamen tube, x 10. Figure 24, top view, x 5

44 / LILIIFLORAE (LILIALES)

LILIACEAE

Scilla violacea Hutch.

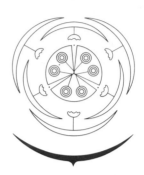

Floral diagram
Note: Since the perianth tube is so inconspicuous at maturity, it is not indicated in the floral diagram.

Floral formula: $* \; P3+3 \; A6 \; \underline{G}(3) \; O6$

Sequence of primordial inception
$P_o 1,2,3$, $P_i 1,2,3$, A_{1-6}, G_{1-3}, O_{1-6}
Note: The inception of the members of each perianth whorl approaches simultaneity in at least some of the buds.

DESCRIPTION OF FLORAL ORGANOGENESIS

The three outer tepal primordia are the first to be initiated on the floral apex. In some cases they seem to appear simultaneously (1), whereas in other cases they arise in clockwise (2) or counterclockwise (4) fashion. Subsequently, the inner tepal primordia become apparent, all at nearly the same time (2-4). One young floral bud was observed which had only five perianth members (5). When the tepal primordia have grown up considerably, the six stamen primordia are initiated, all at about the same time (6, 7). They form a single whorl. During their maturation, they develop an introrse anther on a long filament.

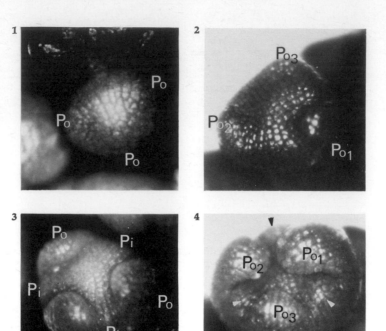

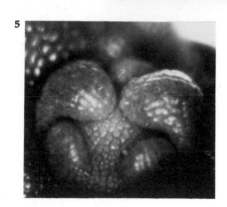

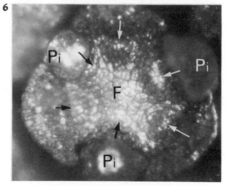

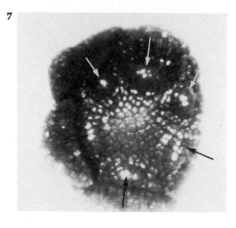

At least in some buds, three stamen primordia on one side mature slightly slower (9-11). For instance, when three of the stamen primordia are distinctly four-lobed, the other three stamen primordia are hardly quadrangular (11). Perhaps the difference in development is due to a difference in primordial inception which might be visible in figures 6 and 7. During the later stages of development some growth occurs between the bases of the young tepals and stamens. As a result, an inconspicuous

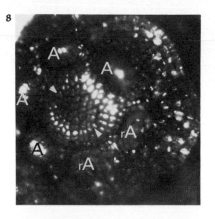

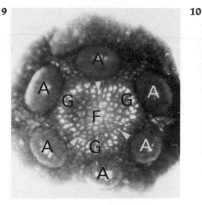

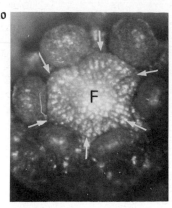

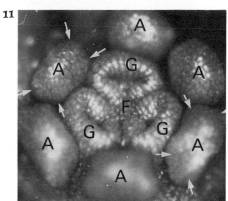

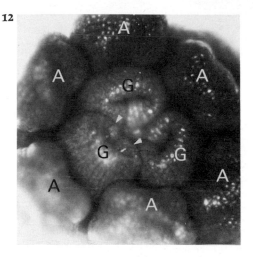

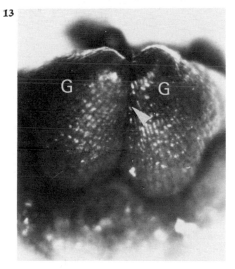

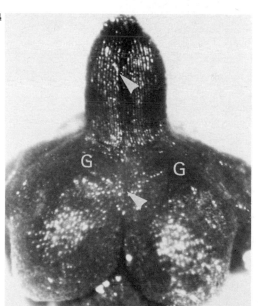

1 Floral bud with incipient outer tepal primordia (P_0). x 146.

-4 Floral buds showing the inception of the outer and inner tepal primordia. Figure 2: clockwise appearance of outer tepals and alternating incipient inner tepal primordia. Figure 4: counterclockwise appearance of outer tepal primordia (P_0) and clockwise inception of inner tepal primordia (see white and black arrowheads). x 146

5 Young bud with five perianth members instead of six. x 146

-7 Floral buds from which outer (6) and inner tepal primordia (7) were removed to show inception of six stamen primordia (see arrows). x 146

FIGURES 8-12. Floral buds from which tepal primordia were removed to show the inception and development of the three gynoecial primordia (G) besides the six stamen primordia (A). x 146

9 Arrowheads point to the region which grows up between the gynoecial primordia.

10 The three gynoecial primordia appear two-lobed (see white arrowheads), perhaps due to presence of the stamen primordia on their flanks. The floral apex (F) is growing upward.

11 The stamen primordia become four-lobed (see white arrows). The margins of the crescent-shaped gynoecial primordia (G) are approaching each other. Also, the floral apex (F) is growing upward.

12 The margins of the gynoecial primordia are almost touching each other (see white arrowheads).

14 Side views of young gynoecia before (13) and after (14) development of styles. Arrowheads point to the line of postgenital fusion between adjacent gynoecial appendages (carpels). Figure 13, x 146. Figure 14, x 85

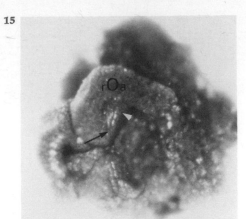

15

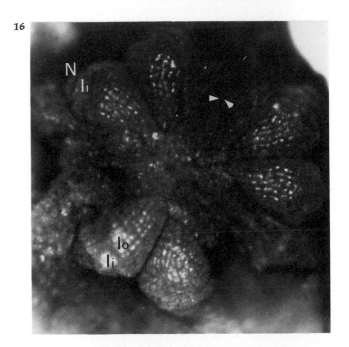

16

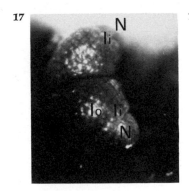

17

18

perianth tube is formed on which the stamens are inserted.

After stamen inception, three slightly crescent-shaped gynoecial primordia appear. At first there is some upgrowth between the margins of these crescent-shaped primordia (8-10), making the gynoecial primordia seem almost like a continuous ridge around the floral apex. But this upgrowth between the gynoecial primordia stops after a short interval. Then the gynoecial primordia become horseshoe-shaped and finally their margins meet in the center of the floral bud (11, 12). At the same time the floral apex grows upward for a short distance. The tip of each gynoecial primordium gradually forms a style, with a terminal stigma, while the lower portion gives rise to the ovary (13, 14). Postgenital fusion occurs from the base of the ovaries to the tip of the styles including the stigmas (14). As a result, one compound pistil is formed.

The ovule primordia are initiated on the inturned margins of the gynoecial primordia (15). There are six ovules and they are all initiated at about the same time. They become bitegmic and anatropous. The inner integument forms first (16, 17).

OTHER AUTHORS

I have found no literature on *Scilla violacea*. Schumann (1890) describes the sequence of tepal inception in *Scilla sibirica* and *Scilla italica*. He notes that the outer tepals arise in succession, the abaxial one being formed last, whereas the inner tepals are formed simultaneously. In *Scilla italica* a prophyll is initiated before the tepals. In *Scilla sibirica* a rudimentary phyllome is formed after the inception of the outer perianth at the outer base of the first formed tepal.

Recently, Greller and Matzke (1970) have described the floral organogenesis of *Lilium tigrinum*.

BIBLIOGRAPHY

Greller, A.M. and Matzke, E.B. 1970. Organogenesis, aestration, and anthesis in the flower of *Lilium tigrinum*. Bot. Gaz. *131*: 304-311.

Schumann, K. 1890. *Neue Untersuchungen über den Blüthenanschluß*. Leipzig: W. Engelmann.

45 / GRAMINALES

GRAMINEAE
(POACEAE)

Hordeum vulgare L. var. Champlain OaC21 (cultivated barley)

Diagram of spikelet

Floral formula: ·|· P2 A3 G̲1 O1
Note: P2 are the two lodicules. Several authors would write G̲(2) because the pistil has two stigmatic lobes.

Sequence of primordial inception
Glumes 1-2, lemma, rachilla, palea, A1-3, P1-2, G$_{girdling}$, G1-2, O1
Note: The lodicules are initiated as indicated above, or perhaps simultaneously with or immediately after gynoecial inception.

DESCRIPTION OF FLORAL ORGANOGENESIS

The first two primordia which appear simultaneously on the spikelet apex are those of the two glumes on the abaxial side (1, black arrows). Subsequently, the primordium of the lemma is initiated (1, Le). This primordium gradually girdles the bud (4-8): its margins grow toward each other, slowly forming a complete ridge (7, 8) around the young spikelet. On the abaxial side of the lemma primordium, the awn appears as a ridge (1, white arrowhead

FIGURES 15-17. Dissected gynoecia showing ovule inception and development. x 146

15 Ovule primordium (see white arrowhead) attached to the margin (see black arrow) of the young ovary whose upper portion was removed (rOa).

16 Six ovule primordia forming the outer integuments (I$_o$) after inception of the inner integuments (I$_i$). Arrowheads point to the postgenital fusion line of adjacent carpel margins.

17 Two ovule primordia, one having only the inner integument (I$_i$), the other with both integuments.

18 Mature flower. x 11

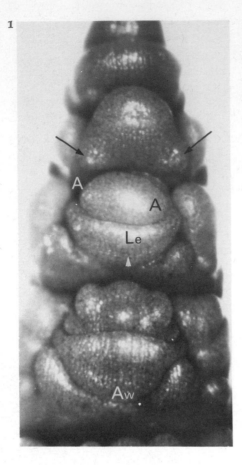

1

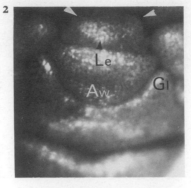

2

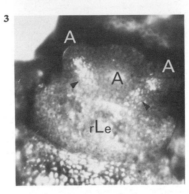

3

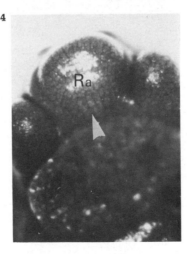

4

and Aw). It grows upward, at maturity projecting above the spikelet (27, 28 Aw). On the adaxial side of the young spikelet, the rachilla forms first as a hump above the extending margins of the lemma primordium (4-8). It does not form further floral primordia but occasionally becomes two-lobed as it matures (15). Above the rachilla, the palea is initiated as a ridge (5, 6) whose lateral portions may grow slightly faster than its median parts (6-8).

Immediately after the inception of the palea, the three stamens appear at about the same time (2-7). They become first two-lobed (10-12), and then four-lobed (13, 14, see arrowheads; 26) as a result of anther formation.

After the inception of the stamen primordia, the floral apex becomes dome-shaped due to growth in height (8, 9). The gynoecium is initiated as a ridge which completely surrounds the floral apex. At first this gynoecial ridge appears quadrangular (10), perhaps due to pressure by the stamen primordia. As it grows upward it becomes circular (11), and subsequently two-lobed (13-17). The two lobes (i.e., the two gynoecial primordia which arise on the original gynoecial primordium) form the two styles with the stigmatic surface consisting of long hairs (24-26). As the styles develop, the ovary closes at the top.

During the inception and early development of the girdling gynoecial primordium, growth extends from its adaxial portion toward the center of the floral bud. In other words, the floral apex (or the center of the floral bud) grows upward in continuity with the adaxial portion of the girdling gynoecial primordium (11, 13, 14). The central portion of this upgrowth gradually forms the ovule, whereas the peripheral portion becomes part of the ovary wall (13-16). As the young ovary grows at its base, the insertion site of the ovule becomes elevated. Thus, in the mature ovary, the ovule is born high up on the ovary wall (24). It is anatropous and bitegmic; the outer integument forms immediately after the inner one (21-23).

The lodicules appear on either side of the abaxial stamen primordium (12). They

---

FIGURES 1-3. Abaxial views of young spikelets. x 146

1 Side view of developing spike with four spikelet buds in different stages of development. On the second bud from top the two glumes are initiated (see black arrows). On the third bud from top, the incipient stamen primordia can be detected (A). The awn primordium (see white arrowhead) is also visible as a ridge on the lemma primordium (Le). On the fourth bud from top the further developed awn primordium (Aw) is visible.

2 Young spikelet with incipient stamen primordia (see arrowheads).

3 Young spikelet with lemma removed (rLe). The three stamen primordia (A) are visible. The incipient lodicule primordia (see black arrowheads) are already present.

FIGURES 4-8. Adaxial views of young spikelets. x 146

4 Appearance of the rachilla primordium (Ra). Also upgrowth between adaxial lemma margins can be seen (see white arrowhead).

5-8 Appearance of palea primordium (Pa, figure 5). Gradual joining of the adaxial lemma margins can be seen (see Le and white arrowhead). Inception of lateral and abaxial stamen primordia (see black arrowheads in figures 5, 6; A in figure 7).

FIGURES 9-17. Top and side views of floral buds. x 146

9 Dome-shaped floral apex (F) and stamen primordia (A).

10 Quadrangular (see arrows) gynoecial primordium. This shape might be due to the pressure of the stamens (A) on the flanks of the young gynoecium.

11 Circular gynoecial primordium (G). On its adaxial side, there is more upgrowth, this being the incipient ovule primordium (see black arrowhead). The young stamens are two-lobed at this stage (see white arrowheads).

12 Abaxial view of floral bud. The lemma is removed, and two lodicule primordia can be seen, slightly below and on either side of the abaxial stamen primordium (see white arrowheads).

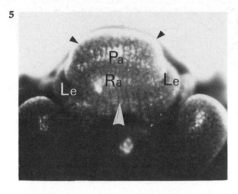

5

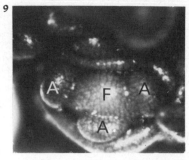

9

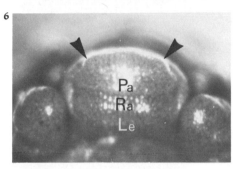

6

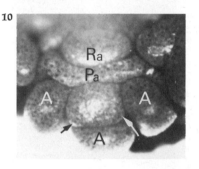

10

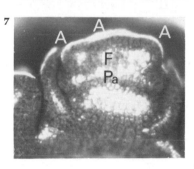

7

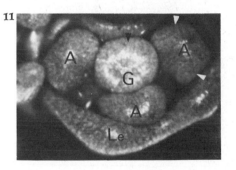

11

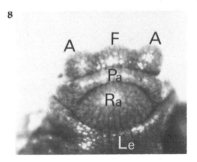

8

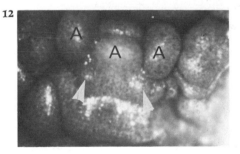

12

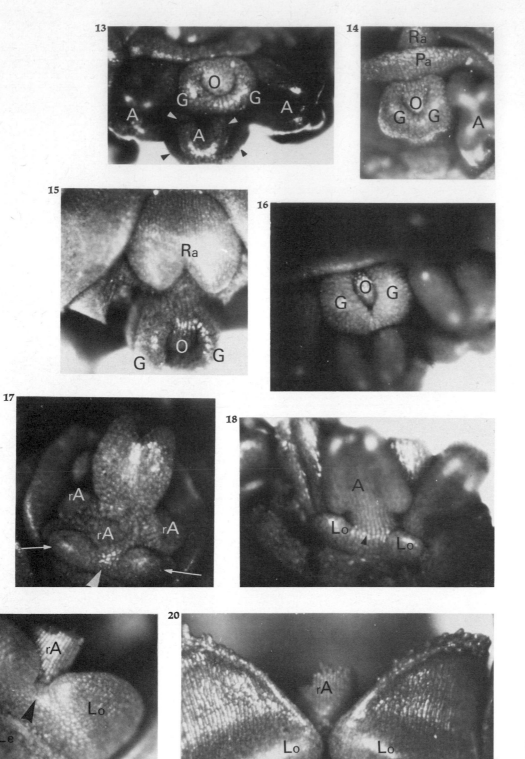

might be initiated before or at the same time as the gynoecial primordium (3, see black arrowheads). Some upgrowth occurs in the region between the lodicule primordia and at the outer base of the abaxial stamen (17-19). However, this growth is so limited that it is no longer noticeable in the mature flower (20). Therefore, it is not indicated in the floral formula and diagram.

OTHER AUTHORS

A number of papers have been published on the floral development of several genera of grasses. Recently, Klaus (1966) and Butzin (1965) have described in detail the spikelet development of barley. All the previous literature on this subject is quoted and discussed by these two authors (e.g., Pankow 1962; Barnard 1955, 1957; Bonnett 1935). Klaus (1966) investigated *Hordeum distichon* L. Butzin (1965) studied both *Hordeum vulgare* L. and *Hordeum distichum* L., and his detailed description of pistil development is based on the latter species. My observations on the development of the androecium and gynoecium agree with both Klaus' and Butzin's results (although both authors use a different and interpretive terminology). There are a few discrepancies with regard to the spikelet development. Butzin (p. 84) mentions the initiation of a rudimentary adaxial glume below the rachilla primordium. I could not detect any such discrete rudiment. But I found a ridge below the rachilla primordium. This ridge is the adaxial extension of the lemma, which completely encircles the spikelet axis (figures 4-8). Klaus (p. 57) mentions that the primordia of the lemma and the glumes toward the abaxial side become visible at the same time. I observed that the lemma is initiated after the glumes.

Mansuri's (1969) study does not add much, if anything, to our knowledge of floral organogenesis.

BIBLIOGRAPHY

Butzin, F. 1965. *Neue Untersuchungen
über die Blüte der Gramineae*. Thesis.
Freie Universität, Berlin.

Klaus, H. 1966. Ontogenetische und histogenetische Untersuchungen an der
Gerste (*Hordeum distichon* L.). Bot. Jb.
85: 45-79.

Mansuri, A.D. 1969. Organogenesis,
growth and development in barley under
varying photoperiodic and vernalisation
treatments. 1. Ontogeny of the shoot
apex. Phyton *26*: 35-46.

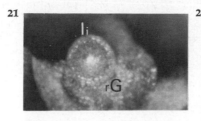

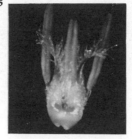

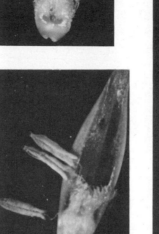

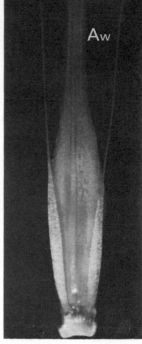

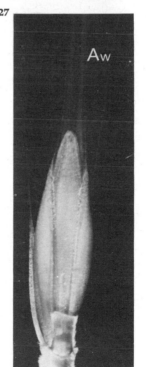

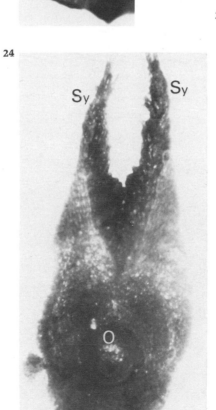

13 Gynoecial ridge with two new primordia (G) on it,
which are the beginnings of the styles. Arrowheads point to the four lobes of the abaxial anther.

-16 Top views of the young gynoecia showing upgrowth of the style primordia (G), enclosing
gradually the ovule. In figure 15 the rachilla is
two-lobed.

17 Side view of a young flower from which the stamens were removed (rA), showing upgrowth of
the gynoecium. The two lodicule primordia (see
white arrows) are visible and the slight upgrowth
(see white arrowhead) of the region between them.

-20 Side views of developing lodicules (Lo). Upgrowth
(see black arrowheads) between the two young
lodicule primordia (Lo). x 146

FIGURES 21-23. Young ovules after the removal of
ovary wall. x 146

21 Only inner integument present (I_i).

-23 Young ovules with both integuments (I_i, I_o, N =
nucellus).

24 Side view of dissected gynoecium, with the styles
(Sy) developing stigmatic hairs. The anatropous
ovule is disclosed (O). x 146

25 Nearly mature flower. x 9

26 Mature spikelet with lemma removed. x 6

27 Adaxial view of mature spikelet (Aw = awn). x 6

28 Abaxial view of mature spikelet. x 6

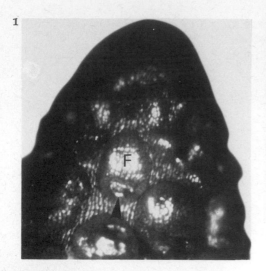

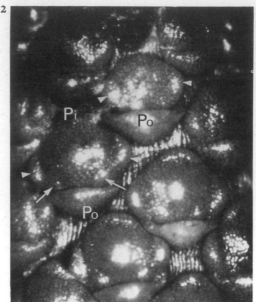

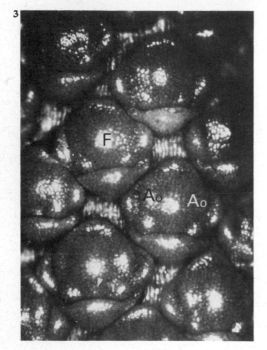

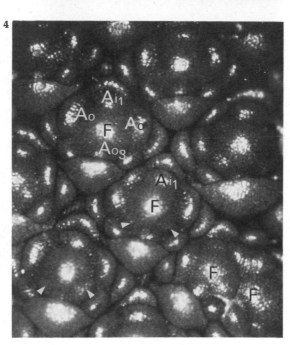

ARACEAE

Acorus calamus L. (sweet flag)

Floral diagram

Floral formula: ✳ P3+3 A3+3 G̲(3) O9-15

Sequence of primordial inception
$P_o1,2-3$, $P_i1,2-3$, $A_o1-2,3$, $A_i1,2-3$, $G1-3$, $O1-9$ to 15

DESCRIPTION OF FLORAL ORGANOGENESIS

The first primordium to appear on the floral apex is the abaxial outer tepal primordium (1). Subsequently, the two outer lateral tepal primordia are initiated (2). Following these, the adaxial inner tepal primordium appears (2), followed shortly by the two inner tepal primordia toward the abaxial side (2).

The first two outer stamen primordia are initiated toward the adaxial side (3). Then, the abaxial stamen primordium of this first whorl arises slightly before the adaxial stamen primordium of the second whorl (3, 4). The last two stamen primordia of the second whorl arise simultaneously on either side of the abaxial stamen primordium of the first whorl (4, 5). As the stamens reach maturity, they form a filament and an introrse anther (18).

Before the inception of the gynoecial primordia, the floral apex enlarges on the flanks. Then three, rarely two, crescent-shaped gynoecial primordia are formed, opposite the stamen primordia of the first whorl (6-12). In the cases when only two gynoecial primordia are formed, these two primordia may be equal in size (11), or one of them may be bigger than the other (10). Immediately after their inception, the crescent-shaped gynoecial primordia grow up in continuity with each other and the floral apex. Thus, a three-locular (three-septate) ovary and three stigmas on three inconspicuous styles develop. Each stigma and style results from appression of the margins of a gynoecial primordium (9-12). Three placentae – one per locule – are initiated on the upper portion of the central column which grows upward in continuity with the septa (13). Three to five ovule primordia appear at the lower extremity of each placenta (14). The ovule primordia form two integuments (16), which become rather elongate (17). Some upgrowth occurs underneath all floral parts with the exception of the base of the locules. This leads

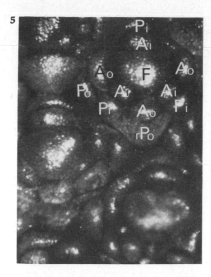
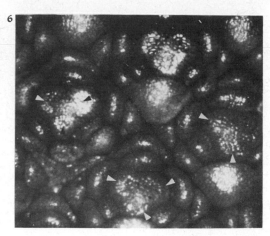
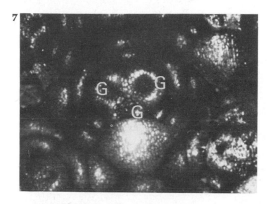

FIGURES 1-5. Side views of portions of young spadices showing floral buds in various stages of development. x 146

1 Appearance of the outer abaxial tepal primordium (see black arrowhead).

2 Top bud with the two lateral outer tepal primordia (see white arrowheads). Bud to the left shows the adaxial inner tepal primordium (P_i) and the two inner tepal primordia toward the abaxial side (see white arrows).

3 The first two stamen primordia (A_o) are visible on the labelled buds. The third stamen primordium of the outer whorl has been initiated in the bud below (see white arrowhead).

4 Second flower bud from top shows the first stamen primordium of the inner whorl (A_{ii}). The bud below shows the simultaneous initiation of the last two inner stamens (see white arrowheads).

5 Floral bud with the abaxial tepal removed (rP_o). The floral apex (F) is triangular as a result of gynoecial inception.

FIGURES 6-11. Floral buds showing stages of gynoecial development. x 146

6-7 Three crescent-shaped gynoecial primordia (see white arrowheads in figure 6 and G in figure 7).

8-9 The margins of the gynoecial primordia (G) are growing toward each other (8), until they finally meet, becoming postgenitally fused (9).

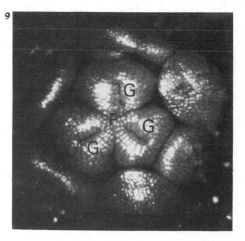

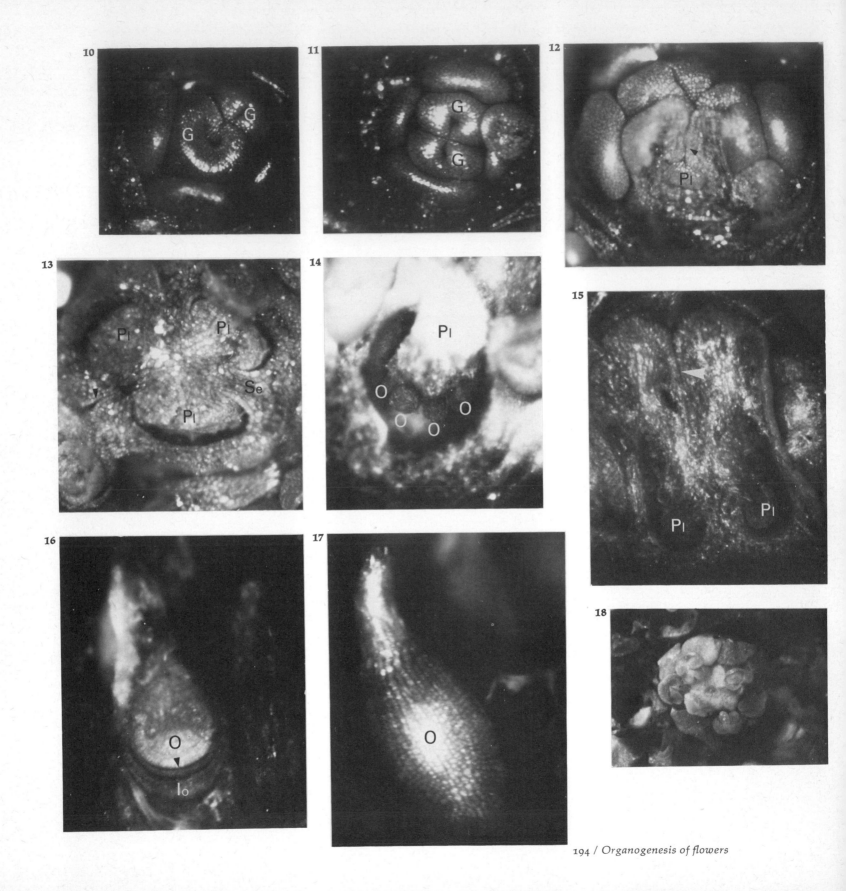

to a slightly epigynous condition which is not indicated in the floral formula because it is hardly noticeable.

OTHER AUTHORS

Payer (1857) describes the floral organogenesis of this species. In contradiction to my observations, he reports that the members of any whorl arise simultaneously; he terms the anthers extrorse; and he mentions that the flower is subtended by a bract.

The results of Eyde *et al.* (1967), who studied the anatomy and morphology of the mature flower, agree with my observations.

BIBLIOGRAPHY

Eyde, R.H., Nicolson, D.H. and Sherwin, P. 1967. A survey of floral anatomy in Araceae. Amer. J. Bot. 54: 478-497.

Payer, J.B. 1857. *Traité de organogénie comparée de la fleur. Texte et Atlas.* Paris: Librairie de Victor Masson.

1 Two crescent-shaped primordia, one of them being much larger than the other (10), or both having equal size (11).
2 Side view of a young pistil from which one gynoecial primordium was removed to disclose the position of one central placental primordium (Pl) laterally on the tip of the central column. x 146
3 Cross-section of young gynoecium. Note a placenta in each of the three locules. Since this is a slightly oblique cross-section, at a higher level free margins of adjacent gynoecial appendages are visible while at a lower level solid septa (Se) can be seen. x 146
4 Side view of one placenta (Pl) with four ovule primordia (O). x 146
5 Longitudinally dissected gynoecium with postgenitally fused margins of the gynoecial appendages at the top (see white arrowhead). Below this fusion line no partition is visible, meaning that upgrowth from the base has taken place. x 146
6 Cross-section through mature ovule (O), with an inner integument (see black arrowhead), and an outer integument (I_o). x 146
7 Mature ovule (O). x 146
8 Mature flower. x 9

47 / PANDANALES

SPARGANIACEAE

Sparganium eurycarpum Engelm.
(broad-fruited bur-reed)

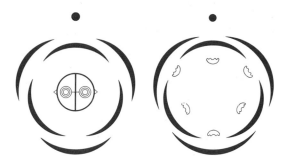

Female flower   Male flower

Floral formula: Floral formula:
$* P_{3+3} \underline{G}(2) O_2$ $* P_{3-6} A_{2-6}$
Note: for variations in floral construction see text.

Sequence of primordial inception
Female flower: $P_o 1,2$-3, $P_i 1,2$-3, $G1$-2, $O1$-2
Male flower: See text

DESCRIPTION OF FLORAL ORGANOGENESIS

Female flower
The first primordium to form on the floral apex is the abaxial outer tepal primordium (which could also be interpreted as the bract primordium) (1). Then, at about the same time, the two outer tepal primordia toward the adaxial side are formed (1). Subsequently, the adaxial inner tepal primordium is initiated (1), followed shortly by the two inner tepal primordia toward the abaxial side, which seem to appear simultaneously (1). In some floral buds all three inner tepal primordia seem to appear simultaneously, or in very rapid succession which approaches simultaneity (2, 3). During their further development, the tepals become scale-like (9).

Before and during the inception of the

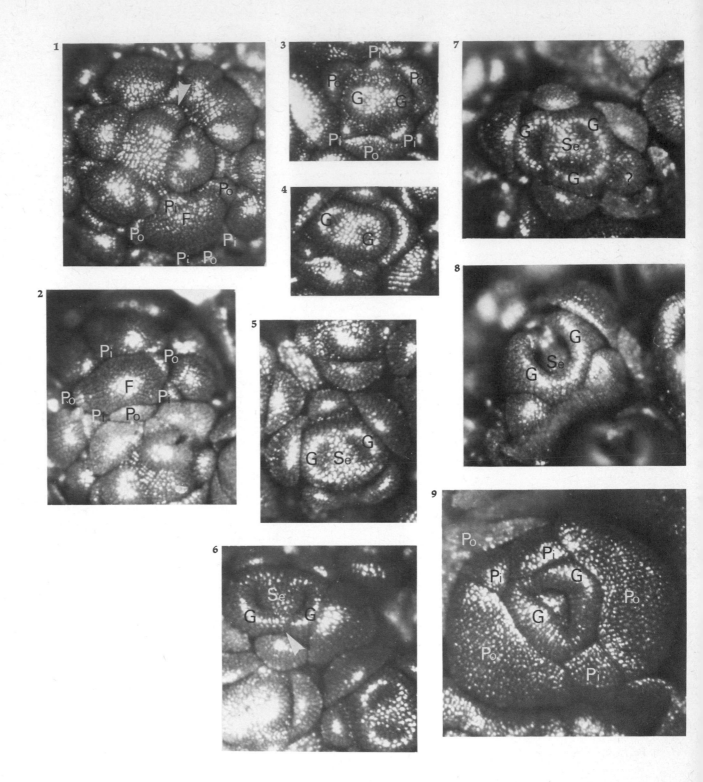

gynoecial primordia, the floral apex broadens in the transversal plane (3). Two (4-6) or rarely three (or four?) (7) crescent-shaped gynoecial primordia are formed. They grow upward in continuity with each other and the floral apex (5-8). In this manner, a two-locular (occasionally three-locular) ovary is formed with the septum in the median plane. The margins of the gynoecial primordia become gradually appressed to each other (12). In this way, the gynoecial primordia develop into the stigmas (10-12, 14). Sometimes, the stigmas are oriented in the median instead of the transversal plane.

Two ovules are formed near the upper end of the central column, i.e., the middle portion of the septum (11-13). They become bitegmic and anatropous, the micropyle pointing toward the septum (13).

Female flower

1 Floral buds with initiating tepals. The large white arrowhead points to the abaxial outer tepal (or bract) primordium of the youngest floral apex. The oldest floral bud shows all outer (P_o) and inner (P_i) tepal primordia. F = floral apex. x 146

2 Floral bud with outer tepal primordia (P_o) and inner (P_i) tepal primordia. F = floral apex. x 146

3 Floral bud with the floral apex enlarging in the transversal plane. Inception of two gynoecial primordia (G). x 146

-9 Top and side views of floral buds with developing gynoecia. Figures 5-6: septum formation (Se) as well as upgrowth from the base of the gynoecia (see white arrowhead in figure 6). Figure 7: three (or four?) gynoecial primordia, instead of the usual two. Figures 8, 9: upgrowth of the gynoecial primordia (G). x 146

12 Young gynoecia with developing stigmas (Si). The margins of each gynoecial primordium fuse post-genitally (see white arrowhead in figure 12) forming a long suture. x 146

13 Longitudinal section through septum, showing the bitegmic ovules and their attachment. I_o = outer integument, I_i = inner integument. x 146

14 Mature female flowers with the two stigmas, ovary and perianth. x 13

10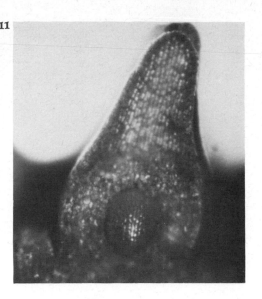

11

12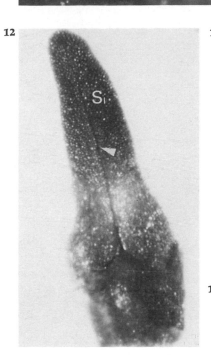

13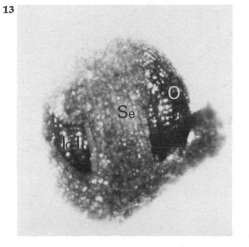

14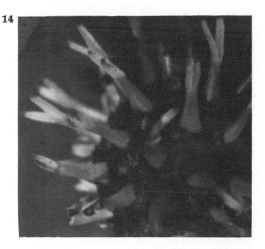

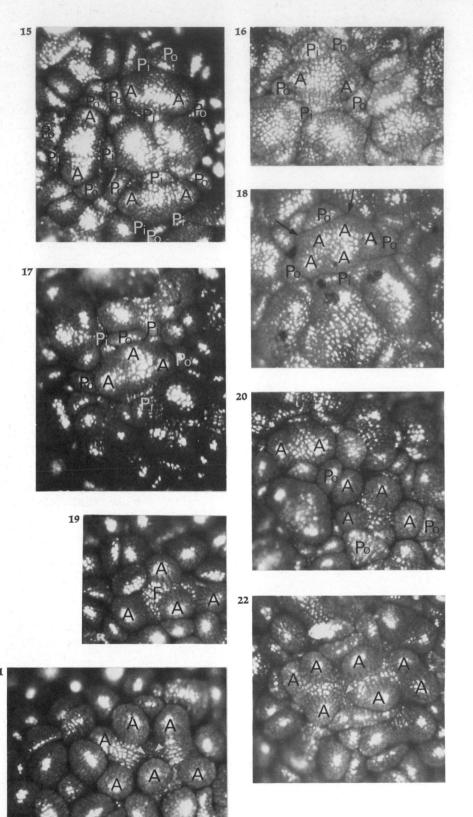

Male flower

There is a great deal of variation in the mode of inception as well as in the number of primordia in the male flowers. The abaxial outer tepal primordium seems to appear first on the floral apex, followed shortly by the two outer tepal primordia toward the adaxial side (15). Variation occurs in the number of inner tepal primordia from zero to three. When there are three inner tepal primordia initiated, the adaxial one appears first, followed closely by the two inner tepal primordia toward the abaxial side (15, 17). Occasionally, only two inner tepal primordia are initiated: one abaxial and one adaxial (16). In other cases, inner tepal primordia are completely missing (20-23), or cannot be distinguished from the young stamen primordia (19, 23, 24).

The number of stamen primordia varies from two to six. Usually, two stamen primordia are initiated first opposite the two outer tepal primordia toward the adaxial side (15, 16, 20). A third outer stamen primordium may or may not be formed. In case it is formed, it will appear opposite the abaxial outer tepal primordium (17). In other buds four (19) or five stamen primordia (18, 20) occur. Floral buds with six stamen primordia also have been found (23, 24). In contrast to the tepals, the stamens elongate considerably. Anthers become extrorse (25).

In a number of cases growth occurred between two adjacent floral buds (21-23). This interfloral growth may become so marked that it is difficult or impossible to delimit individual floral buds.

OTHER AUTHORS

Recently, Müller-Doblies (1969) made a detailed study of the floral development in the genus *Sparganium*. She finds even more variation in the number and position of appendages than I did. In the female flowers she reports a range in tepal number from 1 to 6 (1 to 7 if the bract is included which cannot be distinguished morphologically from the tepals). The most frequent number of tepals was three or four. In the male flowers, she finds a variation between 1 and 6 tepals and 1 and 8 stamens. It is possible that in my material a wider range of variation would have been detected if more material were examined and if more than one population had been used. But the difference may equally well be genetic. Müller-Doblies studied *Sparganium erectum* L., whose relationship to *Sparganium eurycarpum* Engelm. is not fully known. She mentions, however, that the two might be identical at the species level.

With regard to tepal and stamen inception, Müller-Doblies writes (p. 430): 'The perianth segments and the outer stamens are superposed and originate as paired primordia (Goebel, 1911, p. 248-262:

gepaarte Blattanlagen).' Moreover, she mentions that sometimes superposed tepal and stamen primordia seem to originate from one common primordium. I could not detect any such common primordium that gives rise to a stamen-tepal pair. However, there might be a very slight and inconspicuous upgrowth between the tepal primordium and the stamen primordium that is superposed to it.

For more details and a discussion of older literature see the paper by Müller-Doblies (1969).

BIBLIOGRAPHY

Müller-Doblies, U. 1969. Über die Blütenstände und Blüten sowie zur Embryologie von *Sparganium*. Bot. Jahrb. *89*: 359-450.

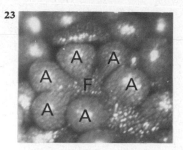

23

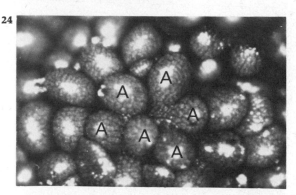

24

25

Male flower

15 Floral buds with initiating outer (P_o) and inner (P_i) tepal primordia. x 146

16 Floral bud with three outer tepal primordia (P_o), two inner tepal primordia (P_i) and two stamen primordia (A). x 146

17 Floral bud with six tepal primordia (P_o, P_i) and three stamen primordia (A). x 146

18 Floral bud with three outer tepal primordia (P_o), one adaxial inner tepal primordium (P_i) and two incipient inner tepal primordia (see black arrows). Five stamen primordia (A) are present. x 146

19 Floral bud with four stamen primordia (A). x 146

20 Floral bud with three outer tepal primordia (P_o) and five stamen primordia (A). This arrangement resembles the bud of figure 18. x 146

22 Continuous buds with six stamen primordia. A slight depression separates the continuous floral apices (see white arrowheads). x 146

23 Floral bud with six stamen primordia and continuity with the adjacent bud. x 146

24 Part of a young male head. x 146

25 Part of male head at maturity. x 13

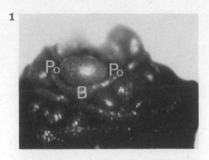

1

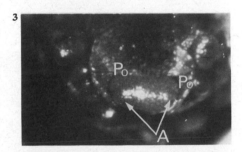

2

3

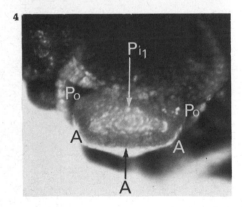

4

48 / CYPERALES

CYPERACEAE

Scirpus validus Vahl.
(strong bulrush)

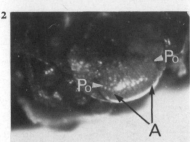

Floral diagram

Floral formula: $\ast$ P3+3 A3 $\underline{G}$(2) O1

Sequence of primordial inception
P_o1-2, A1-2,3 P_i1,2-3, G1-2, O1, P_o3

DESCRIPTION OF FLORAL ORGANOGENESIS

The first two primordia which form on the floral apex are the two outer tepal primordia (1). Subsequently, two lateral stamen primordia appear simultaneously (2, 3). The abaxial stamen primordium appears slightly later or possibly concurrently with the lateral ones (4). As the stamens reach maturity they become two-lobed and introrse (9, 13). The adaxial inner tepal primordium appears at about the same time as the abaxial stamen primordium or slightly before the latter (4). The two inner tepal primordia form next on either side of the abaxial stamen primordium (6, 7).

Before the inception of the gynoecial primordia, the floral apex becomes slightly dome-shaped due to growth in height (5, 6). Then two (rarely three) crescent-shaped gynoecial primordia are formed, opposite the lateral stamen primordia (5, 7-10). In the cases where there are three gynoecial primordia, they would be opposite the stamens (10). There is immediate upgrowth between the margins of these gynoecial primordia (7, 9) as well as upgrowth of the gynoecial primordia, resulting in a cylindrical structure, which will form the ovary (8, 9). As the styles develop, the ovary closes at the top (8, 10).

The floral apex grows up concomitantly with the gynoecial primordia (9), and becomes transformed into the ovule primordium (7, 9, 10). The ovule becomes anatropous and bitegmic (12); the outer integument forms after the inner one (11).

The last primordium to be initiated is the abaxial outer tepal primordium (8), which is inserted at the base of the abaxial stamen primordium.

It must be noted in this flower that, since all the primordia develop very slowly and since the material is extremely small, it is very difficult to determine the sequence of primordial inception without sectioning the material. For instance, the last outer tepal primordium might be initiated before it actually becomes visible externally.

OTHER AUTHORS

Barnard (1957) describes the 'Morphology and Histogenesis of *Scirpus validus* Vahl.' My observations differ in some respects from Barnard's findings. Barnard mentions that 'the anterior perianth member is initiated prior to the posterior member.' If we look at figure 2 of his paper, the section shows a single cell division which he interpreted as the initiation of the abaxial perianth member. Since I have done only dissections of the flower buds, and since the perianth members grow very slowly, this might explain why I did not see the anterior (= abaxial) tepal primordium until after all the other primordia had been initiated. Barnard also mentions that 'the stamens and carpels arise almost simultaneously.' I found all the stamen primordia clearly vis-

ible before the appearance of the gynoecial
primordia. Concerning the gynoecial development, Barnard comments: 'The carpellary tissue originates in periclinal divisions in the hypodermis followed by similar
divisions in the dermatogen. These divisions occur more or less simultaneously in
a ring around the apex and give rise to a circular ridge .... Growth becomes more rapid
at three centers on the encircling ridge and
three peaks are formed. One of these peaks
is anteriorly and two are laterally placed.
Growth of the anteriorly placed peak subsequently ceases, but it continues at the

---

1 Abaxial view of floral bud subtended by a bract
(B) with the two outer lateral tepal primordia
(P_o). x 146

2-3 Adaxial views of young floral buds with the two
lateral stamen (A) and tepal primordia (P_o). x 146

4 Adaxial view of floral bud with the first inner tepal
primordium (P_i1) and the abaxial stamen primordium (see black arrow A). x 246

5 Adaxial views of two floral buds. The bud on the
left with two crescent-shaped gynoecial primordia
(G) with slight upgrowth between their margins
(see arrow). The bud on the right with the dome-
shaped floral apex (see white arrow). x 146

6 Abaxial view of young floral bud with the dome-
shaped floral apex (see black arrow) and with the
two inner tepal primordia (P_i). x 146

7 Top view of floral bud with the inner tepal primordia (P_i) and young ovule primordium (O)
central to the gynoecial primordia. x 146

8 Abaxial view of floral bud and side view of a
more advanced pistil. On the floral bud the last
outer tepal primordium (P_o3) is initiated at the
base of the abaxial stamen primordium (A), also
the style primordia are apparent (see white arrowheads). x 146

9 Adaxial view of young floral bud. The ovule primordium is distinct (O), also the crescent-shaped
gynoecial primordia (G). The two-lobed condition
of the abaxial stamen primordium is apparent (see
white arrowheads). The tepal primordia (P_o, P_i)
are relatively small. x 146

10 Abaxial view of part of an inflorescence. The bud
on the right has the first indication of the two
crescent-shaped gynoecial primordia (see white
arrowhead). At the lower portion of this inflorescence there is a young gynoecium with three developing styles. x 146

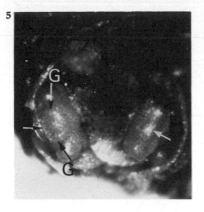

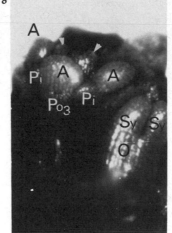

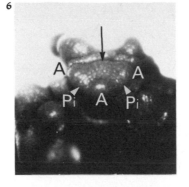

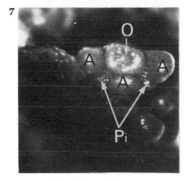

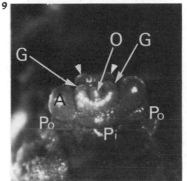

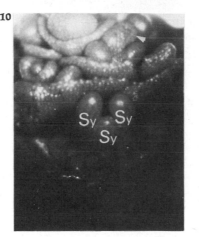

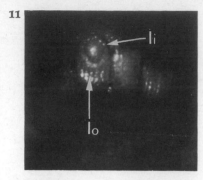

11

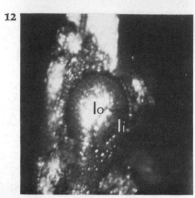

12

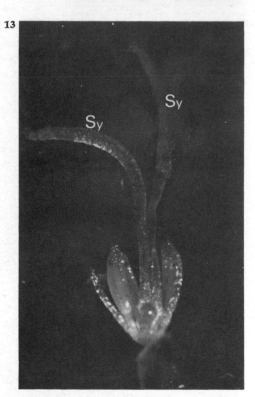

13

two lateral peaks. Each of the lateral peaks develops into a branch of the style.' I found that there are two crescent-shaped gynoecial primordia initiated on the dome-shaped floral apex and that immediately after their initiation there is upgrowth between their margins. Only very rarely did I see three stylar branches at a relatively young stage and at maturity I have only seen two stylar branches. The rest of my observations are in agreement with Barnard's description.

Schultze-Motel (1959) describes the floral development of *Scirpus silvaticus* L. According to his figures 16-18, the abaxial tepal is initiated after the abaxial stamen.

Schumann (1890) makes some observations on the early floral organogenesis of three other species of *Scirpus*. He notes that in these species with three styles the abaxial style (i.e. gynoecial primordium) is initiated later than the two toward the adaxial side and remains shorter for a considerable time. The gynoecial primordia are superposed to the stamen primordia in these species.

BIBLIOGRAPHY

Barnard, C. 1957. Floral Histogenesis in the Monocotyledons. II. The Cyperaceae. Aust. J. Bot. *5* : 115-128.
Schultze-Motel, W. 1959. Entwicklungsgeschichtliche und vergleichend-morphologique Untersuchungen im Blütenbereich der Cyperaceae. Bot. Jb. *78* : 129-170.
Schumann, K. 1890. *Neue Untersuchungen über den Blüthenanschluß.* Leipzig: W. Engelmann.

11- Stages of ovule development after partial removal
12 of the ovary wall. There is an inner integument present in figure 11 (I_i) and the incipient outer integument (I_o), and both integuments are present in figure 12 (I_i, I_o). The anatropous condition of the ovule is evident in figure 12. x 146
13 Mature flower. x 30

49 / CYPERALES

CYPERACEAE

Cyperus esculentus L.
(edible cyperus)

Floral diagram

Floral formula: $* A_3 \underline{G}(3) O_1$

Sequence of primordial inception
$A_{1-2,3}, G_{1-3}, O_1$

DESCRIPTION OF FLORAL ORGANOGENESIS

The first two primordia to appear simultaneously on the floral apex are the two lateral stamen primordia (1). The abaxial stamen primordium forms slightly later (2, 3, 4). As the stamens mature, they become first two-lobed (5), and then four-lobed and introrse.

Before the inception of the gynoecial primordia, the floral apex becomes convex (4). Then, three crescent-shaped gynoecial primordia are formed opposite the stamen primordia (5, 6). There is immediate upgrowth between the margins of these gynoecial primordia, until the original three primordia are hardly distinguishable (7). Upgrowth between the margins of the gynoecial primordia (7) results in a pear-shaped structure (11, 12) which will form the ovary (11, 12, 15). As the styles develop from the gynoecial primordia, the ovary closes at the top (9-12).

The floral apex grows upward at the same time as the gynoecial primordia (7, 10), and becomes transformed into the young ovule primordium (6-10). Two integuments are formed on the ovule primordium, the outer one after the inner one (13). Concomitantly, the ovule becomes anatropous (14).

OTHER AUTHORS

I found no literature on the floral development of *Cyperus esculentus*. However, Schultze-Motel (1959) describes the histogenesis of the flowers of *Cyperus congestus* Vahl. which show the same construction and development as those of *Cyperus esculentus*.

Barnard (1957) and Schultze-Motel (1959) describe the floral histogenesis of *Cyperus eragrostis* Lam. The gynoecium is also trimerous in this species, but in the bisexual flowers only one lateral stamen is present. Barnard, who studied only the female flower of this species, observes cell

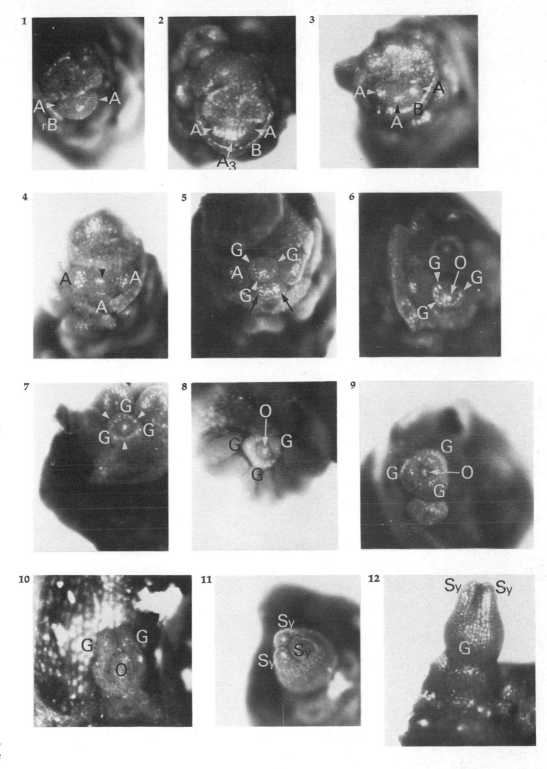

1 Top view of a young floral bud with two lateral stamen primordia (A). The bract is partly removed (rB). x 146

2-3 Top views of floral buds with all three stamen primordia (see A3, A). x 146

4 Top view of floral buds with the dome-shaped floral apex (see black arrowhead). x 146

5 Top view of floral bud with three crescent-shaped gynoecial primordia (G) opposite each stamen primordium. At this stage, the young stamens are slightly two-lobed (see black arrows). x 146

FIGURES 6-7. Top views of floral buds with developing ovules and gynoecia. x 146

6 Three crescent-shaped gynoecial primordia (G) with slight upgrowth between their margins, surrounding a young ovule primordium (O).

7 Upgrowth between the margins (see white arrowheads) of the three gynoecial primordia (G) has made the latter primordia hardly distinguishable.

-12 Stages of pistil development. The young ovule primordium (O) is visible in figures 8, 9. Three young styles (Sy) are formed (11, 12). In figure 10, the ovary wall has been partly removed to show the position of the ovule primordium (O). x 146

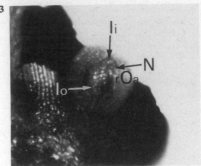

divisions which he thinks might be a 'vestigial expression of a perianth.' These divisions occur after the inception of the pistil.

BIBLIOGRAPHY

Barnard, C. 1957. Histogenesis in Monocotyledons. II. The Cyperaceae. Aust. J. Bot. 5: 115-128.

Schultze-Motel, W. 1959. Entwicklungsgeschichtliche und vergleichend-morphologische Untersuchungen im Blütenbereich der Cyperaceae. Bot. Jb. 78: 129-170.

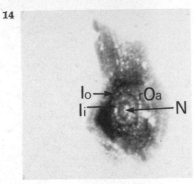

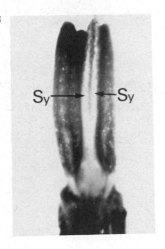

13-14 Side views of dissected ovaries (rOa) with the developing ovules. Figure 13: ovule primordium with inner integument (I_i) showing inception of outer integument (I_o). x 146

15 Nearly mature flower. x 37

50 / MICROSPERMAE

ORCHIDACEAE

Habenaria clavellata (Michx.) Spreng. (green woodland orchis)

Floral diagram

Floral formula: $\cdot | \cdot$ P3+3 A1 $\bar{G}$(3) O∞

Sequence of primordial inception
P_o1-3, P_i1,2-3, A1, G1-3, O1-∞
Note: P_i1 is the lip.

DESCRIPTION OF FLORAL ORGANOGENESIS

The first primordia to appear on the floral apex are the three outer tepal primordia (1). Zygomorphy begins at this early stage (1) due to lateral expansion of the apex. The first inner adaxial tepal primordium is initiated next (3-6). This primordium will develop into the 'lip.' At a later stage of development, it forms an evagination at its base which will grow into a long spur (17, 18, 19). The two lateral inner tepal primordia are formed simultaneously soon after the inception of the 'lip' primordium (4-6).

The single stamen primordium is initiated immediately after the lateral tepal primordia (4-6). As it enlarges, it develops an introrse anther (7, 9-11, 13). On either side of the young stamen two small primordia form, which become two-lobed (10, 12).

(In the literature these structures are referred to as staminodes.)

The gynoecium is initiated as a ridge around the floral apex (8). This ridge shows perhaps more growth in the regions opposite the three outer tepals (8). During further development the abaxial region starts to grow upward and becomes three-lobed (10, 12, 13). Concurrently there is upgrowth below this region and the stamen primordium, leading to the formation of the column (gynandrium). The other two enhanced regions on the gynoecial ridge develop very slowly (12). The ridge itself does not show much further development. The inferior ovary is the result of upgrowth below the ridge and the other floral parts (10, 13, 17). As it develops, three elongated placentae are formed on the same radius as the inner tepals (13, 14). These placentae become highly convoluted and on these convolutions a large number of ovule primordia form (14). The ovules are bitegmic, with the inner integument initiating before the outer (15, 16).

The mature flower becomes resupinated, i.e., the ovary makes a twist of 180°. The floral diagram represents the flower after resupination (18, 19).

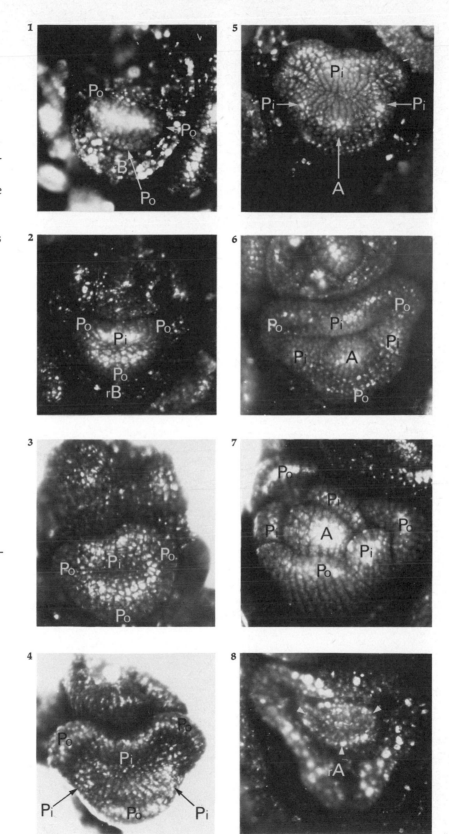

1 Floral apex with the outer three tepal primordia. Bract removed (rB). x 146

3 Floral bud with the outer tepal primordia and the first indication of the inner adaxial tepal primordium which will form the lip (Pᵢ). x 146

4 Floral bud with the incipient lateral inner tepal primordia (Pᵢ). x 146

5 Floral bud with all tepal primordia and the incipient stamen primordium (A). x 146

7 Young buds with primordia of all tepals and stamen. x 146

8 Floral bud with tepal and stamen primordia removed to show the gynoecial ridge around the floral apex. Three regions with perhaps more growth can be seen on this ridge (see white arrowheads). x 146

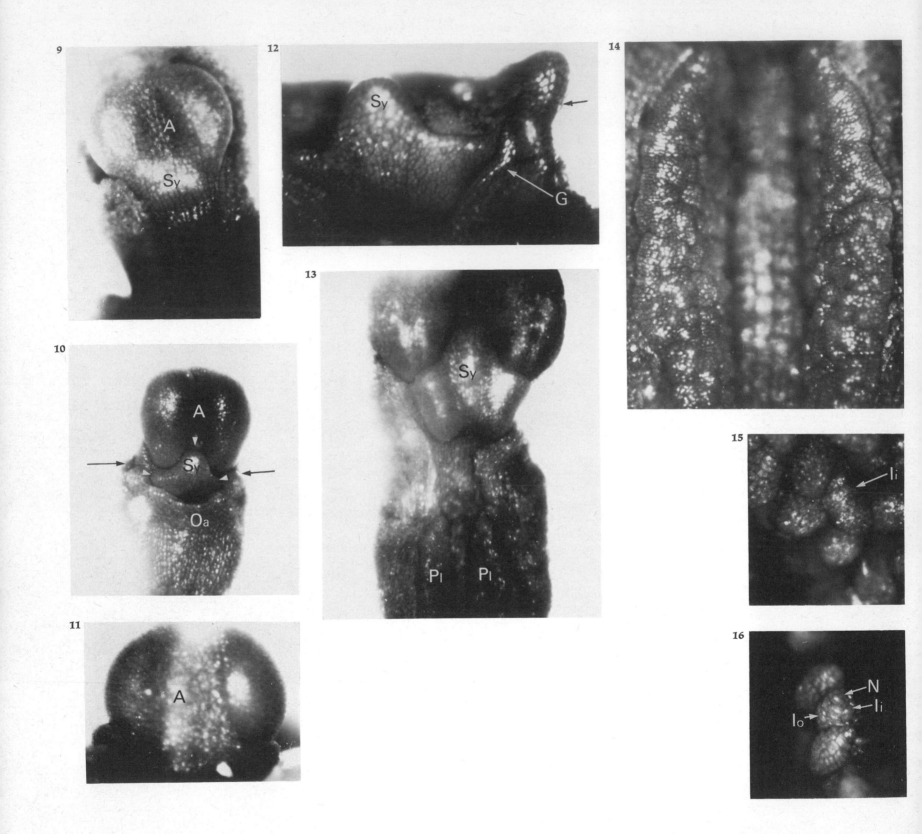

Most of the work that has been done on orchids deals mainly with the mature flower and late stages of development (e.g., Hirmer 1920). I found no literature on the floral development of *Habenaria clavellata*. Pfitzer (1888) describes the floral development of *Orchis morio*. There is agreement between Pfitzer's results and my observations on *Habenaria clavellata*. Pfitzer notes, however, that the lateral outer tepals are initiated before the abaxial outer tepal.

Jeyanayaghy and Rao (1966) describe the 'Flower and seed development in *Bromheadia finlaysoniana*.' They refer to tepal inception, column formation, and gynoecial initiation too briefly to allow fruitful comparison with the floral development of *Habenaria clavellata*. It is to be noted though, that figures 2 and 3 in the above paper agree with my observations.

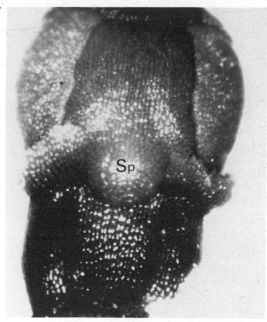

17

Rao (1967) describes 'The flower and seed development in *Arundinia graminifolia*.' He notes that 'among the perianth lobes, the labellum develops last.' In *Habenaria clavellata*, the lip or labellum is initiated before the two lateral inner tepal primordia. The description of column formation is done too briefly by Rao to allow any comparison with the column formation in *H. clavellata*; but the diagrams of the longitudinal sections of *Arundinia graminifolia* (figures 4-7) show a similar column as has been seen in *H. clavellata*. The major portion of Rao's paper deals with microsporogenesis and megasporogenesis in *Arundinia graminifolia*.

Swamy (1948) describes the vascular anatomy in Orchidaceae.

BIBLIOGRAPHY

Hirmer, Max von. 1920. Beiträge zur Organographie der Orchideenblüte. Flora 113: 213-309.
Jeyanayaghy, S. and Rao, A.N. 1966. Flower and seed development in *Bromheadia finlaysoniana*. Bull. Torrey Bot. Club 93: 97-103.
Pfitzer, E. 1888. Untersuchungen über Bau und Entwicklung der Orchideenblüthe. Pringsheims Jahrb. wiss. Bot. 19: 155-177.
Rao, A.N. 1967. Flower and seed development in *Arundinia graminifolia*. Phytomorphology 17: 291-300.
Swamy, B.G.L. 1948. Vascular anatomy of orchid flowers. Bot. Mus. Leafl. Harv. 13: 61-95.

18

19

9 Side view of two-lobed young stamen (A) and incipient style primordium at its base (Sy). x 146
10 Side view of young bud from which the perianth was removed to show the three-lobed style primordium (see white arrowheads) and the young stamen with two-lobed structures (staminode primordia) at either side (see black arrows). x 85
11 Young stamen viewed from the inflorescence axis with a staminode primordium on either side. x 146
12 Side view of young style (Sy), one lateral gynoecial primordium (G) on the gynoecial ridge and the staminode (see black arrow). x 146
13 Floral bud with perianth and part of the ovary wall and the abaxial placenta removed, leaving the two lateral placentae (Pl). The style (Sy) is still present. x 85
14 Side view of two placentae after the removal of the abaxial portion of the ovary wall. Profuse convolutions develop on the placentae before the inception of the ovules. x 146
16 Ovule primordia after the inception of the inner integument (15, I_i), and outer integument (16, I_i, I_o). x 146
17 Adaxial view of flower bud showing the initiation of the spur (Sp) at the base of the inner tepal. x 146
18 Mature flower after resupination. x 5
19 Mature flower. Only part of the spur (Sp) and ovary (G) is shown. x 18

This book

was designed by

ANTJE LINGNER

under the direction of

ALLAN FLEMING

and was printed by

University of

Toronto

Press